Eco*Ulysses*

STUDIES IN LITERATURE, CULTURE, AND THE ENVIRONMENT

Edited by Hannes Bergthaller, Gabriele Dürbeck, Robert Emmett, Serenella Iovino, Ulrike Plath

Editorial Board:
Stefania Barca (University of Coimbra, Portugal)
Axel Goodbody (University of Bath, UK)
Isabel Hoving (Leiden University, The Netherlands)
Dolly Jørgensen (Umeå University, Sweden)
Timo Maran (University of Tartu, Estonia)
Serpil Oppermann (Hacettepe University, Ankara, Turkey)
Dana Phillips (Towson University, Baltimore, USA)
Stephanie Posthumus (McGill University, Montreal, Canada)
Christiane Solte-Gresser (Saarland University, Saarbrücken, Germany)
Keijiro Suga (Meiji University, Tokyo, Japan)
Pasquale Verdicchio (University of California, San Diego, USA)
Berbeli Wanning (University of Siegen, Germany)
Sabine Wilke (University of Washington, Seattle, USA)
Hubert Zapf (University of Augsburg, Germany)
Evi Zemanek (University of Freiburg, Germany)

VOLUME 4

Yi-Peng Lai

Eco*Ulysses*

Nature, Nation, Consumption

Bibliographic Information published by the Deutsche Nationalbibliothek
The Deutsche Nationalbibliothek lists this publication in the Deutsche
Nationalbibliografie; detailed bibliographic data is available in the internet at
http://dnb.d-nb.de.

Library of Congress Cataloging-in-Publication Data
A CIP catalog record for this book has been applied for at the Library of Congress.

Cover image: © Yi-Peng Lai

ISSN 2365-645X
ISBN 978-3-631-74403-1 (Print)
E-ISBN 978-3-631-76224-0 (E-PDF)
E-ISBN 978-3-631-76225-7 (EPUB)
E-ISBN 978-3-631-76226-4 (MOBI)
DOI 10.3726/b14423

© Peter Lang GmbH
Internationaler Verlag der Wissenschaften
Berlin 2018
All rights reserved.

Peter Lang – Berlin · Bern · Bruxelles · New York ·
Oxford · Warszawa · Wien

All parts of this publication are protected by copyright. Any
utilisation outside the strict limits of the copyright law, without
the permission of the publisher, is forbidden and liable to
prosecution. This applies in particular to reproductions,
translations, microfilming, and storage and processing in
electronic retrieval systems.

This publication has been peer reviewed.

www.peterlang.com

For my parents, who have taught me how to grow things.

At Fredericksburg, as they walked through the grounds, Vinding asked, 'Do you like flowers, Mr. Joyce?' 'No. I love plants, green growing things, trees and grass. Flowers annoy me.'
— Richard Ellmann, *James Joyce,* p. 694

Acknowledgment

Numerous people have contributed to the birthing of this project, without whom it would not be in existence. Above all others I owe my gratitude to Professor Brian G. Caraher, my Ph.D. supervisor at Queen's University, Belfast, for having patiently and faithfully guided me and encouraged me over the course of the birthing of this current work. I owe my thanks to those who have greatly inspired me throughout this journey: to Professor Lin Yu-chen, for introducing me to Joyce and continuing to walk alongside me like a mother-mentor; to Professor Michael Levenson, for introducing me to *Ulysses*, inspiring me to undertake this project, and cheering me on; to Professor Anne Fogarty, Dr. Eamonn Hughes, and Dr. David Dwan, for kindly reading through my materials and giving me invaluable feedbacks for my immature project at various stages of my research.

I am much indebted to the generous two-year overseas scholarship granted by Ministry of Education of Taiwan, the Zürich James Joyce Foundation Scholarship, which gave me an opportunity to conduct my research in Zürich for two months, and the numerous fundings kindly granted by UCD James Joyce Research Centre, Dublin James Joyce Summer School, Trieste Joyce School, Keogh-Naughton Notre-Dame Centre of Irish Studies, and ASLE (Association for the Study of Literature and Environment), for encouraging me and permitting me to participate in different thought-provoking academic activities during my years of research for this project.

Earlier versions of this book first appeared elsewhere: a version of chapter 1 appeared in *Polymorphic Joyce: Joyce Studies in Italy* 12 (Roma 2012), and a version of chapter 3 was previously published in *Eco-Joyce: The Environmental Imagination of James Joyce* (Cork: Cork University Press, 2014). I would like to thank the above publishers for their permissions to reprint parts of these earlier essays. I'd also like to thank Hannes Bergthaller for his kind support and encouragement as series editor of the "Studies in Literature, Culture, and the Environment" series, Michael Rücker, Magda Kalita and others at Peter Lang for their assistance and patience, and two anonymous readers for their insightful comments. To help launch the publication of this book, the College of Liberal Arts at National Sun Yat-sen University also granted me a reimbursement, for which I'm much obliged.

Finally, to those whose love has made me who I now am: thank you.

Contents

Abbreviations 13

INTRODUCTION: Toward an Eco-political
Reading of *Ulysses* 15

PART I NATURE, LAND, CONSUMPTION

1 GARDENS: Irish Land Reform, Garden Cities and
   the Politics of Gardening 31

2 WASTE: Joyce's "cloacal obsession" and the Eco-politics
   of Waste 69

PART II NATURE, NATION, ECOSYSTEM

3 TREES: History, Performance, and the Viconian
   Politics of the Forest 101

4 PASTORAL: Social Languages, Politics,
   Nature and Nation 141

CONCLUSION: The Journey and Beyond 171

List of Figures 173

Bibliography 175

Index 189

Abbreviations

FW	Joyce, James. *Finnegans Wake*. New York: Penguin, 2000.
JJ	Ellmann, Richard. *James Joyce*. New and revised edition. Oxford: Oxford University Press, 1982.
LI	Joyce, James. *Letters of James Joyce: Volume I*. Ed. Stuart Gilbert. New York: Viking Press, 1966.
LII	Joyce, James. *Letters of James Joyce: Volume II*. Ed. Richard Ellmann. New York: Viking Press, 1966.
LIII	Joyce, James. *Letters of James Joyce: Volume I*. Ed. Stuart Gilbert. New York: Viking Press, 1966.
NS	Vico, Giambattista. *The New Science of Giambattista Vico: unabridged translation of the third edition (1744) with the addition of "Practice of the New Science."* Trans, by Thomas Goddard Bergin and Max Harold Fisch. Ithaca: Cornell University Press, 1984.
OCPW	Joyce, James. *Occasional, Critical, and Political Writing*. Oxford University Press, 2008.
U	Joyce, James. *Ulysses*. Ed. Hans Walter Gabler. New York: Vintage, 1993.
U-G	Joyce, James. *Ulysses: A Critical and Synoptic Edition*. Volume I-III. By Hans Walter Gabler, Wolfhard Steppe and Claus Melchior. New York: Garland, 1986.

INTRODUCTION
Toward an Eco-political Reading of *Ulysses*

> *In one sense, the Irish problem has persisted because of the power of geographical images over men's minds.*
> —Oliver Macdonagh, *States of Mind*, p. 15

In praising the beauty of nature, poets and writers throughout time have never reached the limits of their language. William Wordsworth "could not but be gay/In such a jocund company"[1] of the daffodils, while John Keats admires the never-dying "poetry of earth."[2] These Romantic verses, expressing naïve or sentimental admiration for an ancient pastoral tradition since the time of Theocritus, share a political "anti-urbanism" with its nostalgia for a simpler rural life. A century later, such evocation of nature is married with more political anxieties in the Irish context. W. B. Yeats, in "The Lake Isle of Innisfree," has the narrator claiming he "will arise and go now, and go to Innisfree,/And a small cabin build there, of clay and wattles made;/Nine bean-rows will [he] have there, a hive for the honey-bee;/And [will live] alone in the bee-loud glade." In "Easter, 1916," the ever-changing nature of "birds," "stream," and "shadow of cloud" elicits Yeats' patriotic concerns for Ireland. In *The Land of Heart's Desire*, the angelic presence of death, emerging with a song of nature, arouses the often depressive agricultural anxiety of Irish rurality.

This conflict between urban and rural, and between England and Ireland, induces modern anxieties for both national and individual consciousnesses. By imagining nature beyond the confines of Ireland's insular geography, geographical formations and ecological writings together posit a struggling force against the urban, the civilized, the imperial, as well as the constructed ideology. In response to such modern development, Raymond Williams' canonical study *The Country and the City* was probably one of the first and the most influential works devoted to the question of urbanization and the dynamics of such conflict in the English literary tradition after industrialization. As McDowell notes accordingly, "[l]andscape writing is permanently embroiled in this struggle. Typically a speaking voice goes out to encounter the landscape and all its elements, an 'on the road' pattern from at least the *Odyssey* onward"[3].

The spatial dynamics of *Ulysses*, with the novel modeled after the mythological journey of *Odyssey*, thus conjure up topographical and geographical concerns as part of the epic backdrop. Beyond the "spatially real"[4] urban setting of Dublin throughout *Ulysses*, the ecological imagining of rurality occasionally intrudes, offering an opposite type of locality to the metropolis of the capital. Such an urban-rural complex derives not merely from a colonial uneasiness within the cityscape, but also from the unsatisfactory desire for the Irish countryside as a result of agricultural[5] development and Ireland's complicated history concerning the land.

The problem of Ireland is the problem of land. As history unfolds, Irish nationalism and political movements initiate from the awakening of agrarian consciousness. The Great Famine and post-famine depressions from the mid-19th century had forced the already problematic landlordism to extremes, and peasant attachment to the land grew into nationalism as an inevitable result of the country's colonial situation. Under the aggressive promotion of the Irish National Land League and Charles Stewart Parnell, Ireland went through several stages of land reform, which in turn led to the Land War (1879-1882), contributed to the growing Irish ownership of land and rural housings, and paved the way for the successive Home Rule Bills (1886, 1893, 1912-14).[6]

In his groundbreaking study of Irish culture, *Heathcliff and the Great Hunger* (1995), Terry Eagleton persuasively points out the political dynamics of the cultural imagination of Nature in the Irish context:

> [...] throughout the British nineteenth century, a chronically idealizing, aestheticizing discourse of both Nature and society was secretly at loggerheads with an altogether more gross, materialist language, heavy with biological ballast and grotesquely bereft of 'culture'. This was the language of bourgeois political economy, which speaks of men and women as laboring instruments and fertilizing mechanisms in a kind of savage Swiftian reduction utterly out of key with the legitimating idiom of cultural idealism. The problem for the ruling British order was that this brutally practical discourse threatened to demystify its own idealizations; and this, in part, reflects a conflict between a kind of language organic to the industrial middle class, and one largely inherited by it from its patrician predecessors. Ireland, in this as in other ways, then comes to figure as the monstrous unconscious of the metropolitan society, the secret materialist history of endemically ideal England. It incarnates [...] the Tennysonian nightmare of a Nature red in tooth and claws, obdurately resistant to refinement.[7]

Indeed, as Eagleton indicates, it would seem more probable that, in Ireland, a landscape marked by the historical scars of famine, deprivation and dispossession, can never present itself to human perception with quite "the rococo charm of a Keats, the sublimity of a Wordsworth or the assured sense of proprietorship of an Austen"[8]. Nature in Ireland is instead moralized and sexualized, and, when transcendentalized, highly political in its portrayals. *Heathcliff and the Great Hunger* takes on much of the emerging critical colonial discourses in

Irish studies and presents a refreshing study of Nature and Irish culture from a Marxist perspective. Two years after the publication of *Heathcliff and the Great Hunger*, the collection *Nature in Ireland: A Scientific and Cultural History* (1997) was published by the Lilliput Press. Edited and compiled by John Wilson Foster, this sizeable collection was possibly the first interdisciplinary book project attempting to bridge scientific studies and cultural studies of Ireland. As Foster comments in the Preface to this collection, one of the tasks *Nature in Ireland* hopes to perform is "to rehouse natural history in Irish culture, from which it has effectively been evicted for a century or so," and it can be done only "when the systematic study of nature is put inside a cultural and intellectual setting broader than the history of scientific disciplines"[9]. This richly interdisciplinary volume contains chapters devoted to specific areas of study such as geology, woodland, botany, entomology, mammalogy, etc., as well as natural history, demesnes, naming, education, the culture of nature, and so on. Such a broad range of topics and contributors allows space for conversation: though few articles in this collection manage to make it very far in their so-called interdisciplinary approach, the wide range of academic voices on nature and Ireland in the very same volume opens up the possibility for the convergence of ideas and perspectives on nature and culture and is in itself revolutionary. Although the word "ecocriticism" is never mentioned in this collection, it has the same aim — the combined study of nature and culture — that "ecocriticism" as an interdisciplinary area of study purposes to accomplish.

In the introduction to *The Ecocriticism Reader: Landmarks in Literary Ecology* (1996), the first anthology of classical and cutting-edge writings in the emerging field of literary ecology, Cheryll Glotfelty defines ecocriticism as an area of study that "takes as its subject the interconnections between nature and culture, specifically the cultural artefacts of language and literature. As a critical stance, it has one foot in literature and the other on land; as a theoretical discourse, it negotiates between the human and the non-human"[10]. Although ecocriticism as a discipline was not recognised as a theoretical approach to cultural studies until the publication of *The Ecocriticism Reader*, the term "ecocriticism" was first introduced almost two decades earlier by William Rueckert in his 1978 article "Literary and Ecology: An Experiment in Ecocriticism" with an aim to "join literature to ecology"[11]. *The Ecocriticism Reader* collection, being the first of its kind in literary ecology, along with the founding of the Association for the Study of Literature and the Environment (ASLE) and the association's allied journal *Interdisciplinary Studies in Literature and the Environment* (*ISLE*), led to ecocriticism's subsequent rapid growth. However, since the 2000s, in response to the challenging voices demanding definitions of terms like "nature," "ecology," and "culture," a second-wave movement started to emerge. Lawrence Buell, a prominent eco-critical voice in both first-wave and second-wave movements, comments thus:

> Although I believed then and continue to believe that the literatures of nature *do* bear important witness against 'the arrogance of humanism' (Ehrenfeld 1978), I found myself agreeing with those who thought the concentration on 'environment' as 'nature' and on nature writing as the most representative environmental genre were too restrictive, and that a mature environmental aesthetics — or ethics, or politics — must take into account the interpenetration of metropolis and outback of anthropocentric as well as biocentric concerns.[12]

This opens up the ground for ecocriticism: instead of focusing merely on ecologically oriented criticism of texts, the study of ecocriticism could now be referred to as environmental criticism, cultural ecology, eco-fiction, literary ecology, ecotheory, ecofeminism, ecopolitics, or literary environmental studies, depending on the subject and the direction of the critic. Together with Lawrence Buell's *The Future of Environmental Criticism* (2005), Greg Garrard's *Ecocriticism* (2004) offers a wide-ranging and systematic study of the discourse, expanding the terrain of ecocriticism from the previously predominant North American nature writing in the first-wave movement to the more recent attention to literatures of diverse topics dealing with gender, race, urbanism, contested landscapes in colonial zones, Darwinism, environmental justice, political ecology, as well as issues of place and cultural geography. Instead of working on relatively recent primary environmental texts as most first-wave eco-critics did, second-wave eco-critics can extend their research to literatures of nearly any period "as a way of helping us understand the emergence of our present environmental crisis"[13]. It is with the rise of second-wave ecocriticism and its attention to diverse discourses on colonial and cultural studies that connections between Irish studies and ecocriticism began to flourish.

Gerry Smyth's "'Shite and Sheep': An Ecocritical Perspective on Two Recent Irish Novels" (2000) and *Space and the Irish Cultural Imagination* (2001) were the first scholarly publications that put discussions of Irish culture in conversation with ecocriticism. Smyth points out that "Irish studies and ecocriticism will have a lot to say to each other.... Geographical peculiarity and historical discontinuity have produced a situation in Ireland in which questions concerning space, landscape, locality, gender, urban and rural experience and nature have become central to both the cultural and the critical imagination"[14]. Following Smyth's pioneering text one may see a blossoming of ecocriticism in Irish studies, including publications such as Tim Wenzell's *Emerald Green: An Ecocritical*

Study of Irish Literature (2009), Christine Cusick's edited collection *Out of the Earth: Ecocritical Readings of Irish Texts* (2010), and Eamonn Wall's *Writing the Irish West: Ecologies and Traditions* (2011). In the meantime, not necessarily associating themselves with ecocriticism *per se*, scholars of Irish studies have published on nature, the environment and cultural history related to Ireland. These researches have provided a means of conversation between nature and culture, and between literature and environment, and has greatly assisted academic discussions of literary ecology in the Irish context.

However, despite the growth of ecocriticism in Irish studies, there remain a limited number of publications on Joyce and ecocriticism. In *James Joyce in Context*, the critical overview of Joycean scholarship released in 2009, Sean Latham, editor of the *James Joyce Quarterly*, signals an ecocritical approach to Joyce's works as an unexplored field in spite of the recently vibrant ecocriticism movement. In his article "Twenty-first-century critical context," he indicates that "despite all the work that has been done on theorizing Joyce's modernity, missing from almost all the accounts is an awareness of ecology and a critical attentiveness of Joyce's own theory of nature"[15]. Since then, there has been an emerging awareness of ecocriticism in Joycean studies. At the International James Joyce Colloquium which took place in Dublin in 2011, two panels were devoted exclusively to ecocriticism and Joyce. In 2015, The Florida James Joyce series published Alison Lacivita's monograph *The Ecology of* Finnegans Wake (Gainesville: University Press of Florida, 2015), a groundbreaking ecocritical study on *Finnegans Wake*. Another milestone was the publication of the first ecocritical collection of Joycean criticism, *Eco-Joyce: the Environmental Imagination of James Joyce* (Cork: Cork University Press, 2014), which extends the range of ecocritical conversations to various Joycean works from his journalistic writings to *Finnegans Wake*. Divided into sections on "Nature and Environmental Consciousness in Joyce's Fiction," "Joyce and the Urban Environment," and "Joyce, Somatic Ecology and the Body," the collection boasts articles addressing a diversity of issues outlined by critics — be they emerging ecocritics or seasoned Joyceans taking on ecocritical perspectives. A few articles in this collection, such as Cheryl Temple Herr's "Joyce and Everynight," Margot Norris's "Negative Ecocritical Visions in 'Wandering Rocks,'" Greg Winston's "'Aquacities of Thought and Language': The Political Ecology in *Ulysses*," and Eugene O'Brien's "'Can excrement be art ... if not, why not?' Joyce's aesthetic theory and the Flux of Consciousness," specifically address the dynamics between languages, ecocriticism and "dark ecology," a term coined and proposed by Timothy Morton. Some other essays, such as Erin Walsh's "Word and World: The Ecology of the Pun in *Finnegans Wake*" and Gerry Leonard's "Ineluctable Modality of the Visible: 'Nature' and Spectacle in 'Proteus,'" explore the ecology of languages in *Finnegans Wake* and *Ulysses* respectively, thus bringing together discourses on modernism and emerging ecocritical voices.

Since its publication, this volume has received reviews from at least six international peer-reviewed journals, including *Textual Practice*, *ISLE* (*Interdisciplinary Study of Literature and Environment*), *Estudios Irlandeses*, *BREAC* (University of Notre Dame), *Green Letters* and *Irish Studies Review*. Different critics have commented on *Eco-Joyce* as "the initial foray of a critical movement that will saturate Joycean scholarship in the years to come"[16], and have noted how essays in this collection "offer an important foray into limning that ecopoesis — an undertaking which has crucial implications not only for Joyce criticism and modernist studies, but for the enterprise of understanding imaginative literature's role in our environmental future"[17]. Martin Ryle's review for *Textual Practice*, among others, interestingly signals the anthropocentric context of social ecology and its relation with Joyce's writings:

> Joyce's fiction celebrates kinds of pleasure and conviviality that do not require wealth and immoderate material consumption; some of them depend on the 'backwardness' of a Dublin whose slum housing and endemic poverty were notorious but whose streets in those days were for people, not cars. In this celebration, his works speak to the strand of European politics traceable to the green and eco-socialist movement of the 1970s: a formative context for Raymond Williams' thinking on literature and ecology, and one of ecocriticism's points of origin. Human flourishing, in a sustainable society, was and is the (far-off) goal of that green politics, and few writers have more to say than Joyce about how we can and might flourish.[18]

Ryle's arguments offer an interesting, relevant context for this current study. Whereas studies of cultural politics, revivalist politics, and genetic criticism have been blossoming in Joycean scholarship in recent decades, their association with ecocriticism has yet to be made. Ryle's critical response leads us to ask: what is Joyce's perspective on ecology as a system and as a branch of science? How do his concerns with urban problems and environmental issues inform his writing of modernity? And how do historiography, cultural politics and revivalist politics come into play alongside the writing of nature in *Ulysses*? These are some of the issues I will attempt to address in the pages to follow.

Eco*Ulysses*

This project started with questions of a first-time reader of *Ulysses* as simple as "Why Nature?" "Why trees?" "Why garbage?" Since then, it has gone on to develop itself as the research unravels new ways in which languages and history work to give new meanings to diverse expressions of Nature in their distinctive (eco-)political context. This study thus focuses on the relationship between environment, history, politics, and languages in James Joyce's *Ulysses*. Delving into different aspects of Joyce's use of nature and rhetorical discourses in orchestrating a specific dynamic of eco-politics, it adopts an interdisciplinary approach that includes cultural politics, historiographical poetics, genetic criticism along with the close reading of the text. I also examine how expressions and figurations of Nature—imaginary or re-imagined—at times become audible voices to narrate, or even proclaim, a collective or discrepant colonial Irish experience. Owing to Joyce's use of languages in his composition of *Ulysses* and my use of genetic materials for this research, I have found current ecocritical theories wanting in supporting my discussion of the languages of the eco-politics in *Ulysses*. Hence, inspired by my intent to unravel the simple "why?" questions regarding the distinctive topic in each chapter, I have chosen, instead of undertaking ongoing ecocritical discourses, to approach my subjects with an original interdisciplinary study involving natural history, cultural politics, historiography, colonial discourse, genetic studies, and even architectural history.

The chapters in this book are divided into two major sections: the first section, entitled "Nature, Land, and Consumption," addresses the environmental questions of land and consumption through discussions on co-operative politics, the garden city movement, and the eco-politics of waste. Reading into "Ithaca" with parallel passages drawn from "Circe," in the first chapter in this section I look at Leopold Bloom's imagined identity as a farmer transforming into (and "ascending" towards) a socialist statesman, and how the Irish history of land-politics underlies the strains in passages from both episodes. To further my discussion of "Bloom of Flowerville," I ponder over Bloom's desirable farmer figure as a consumerist response to commodity culture against the rise of property-owning farmers who benefited from land reforms. I argue that "Bloom of Flowerville" may be the Joycean representative of a rising farmer class, one that is capable of entrepreneurial projects and engages fervently in political reforms. I also trace the sociopolitical context of the *Irish Homestead*, a propagandistic periodical aiming to educate newly risen farmers with mechanic appliances, and examine how the journal might appeal to Bloom despite its dismissal by Stephen Dedalus.

Proceeding from there, this chapter further pursues a socio-political reading of the domestic imagination of the "Flowerville" (Flower Village) ideal in James Joyce's *Ulysses* against the rise of the "garden suburb" movement in early twentieth-century Ireland. Initiated by Sir Ebenezer Howard's *Garden Cities of To-morrow* (1902), the garden-city urban planning scheme was first popularised in England, and soon started to influence the design of the ideal Irish community, especially during the period in which the Home Rule sentiment was at its peak. However, it was also the time in which Irish Revivalists engaged in fervent activities to resist constitutional nationalism. Whereas these Revivalists promoted independent national industries against monopolizing British capital in the market, their anti-British, and hence anti-materialistic and anti-capitalist, attitude is ambivalently reflected in the Anglicised garden-city programme which they, as exemplarily voiced by AE in the *Irish Homestead*, propagandized. Through a close reading of "Bloom of Flowerville," this chapter attempts to unravel the entangled politics between the (Anglo-) Irish Revival and its corresponding Co-operative movement, the British urban planning movement, and relevant discourses of Irish cultural nationalism. Taking the economic, political and architectural histories of modern Ireland into consideration, I propose a socio-political interpretation of Bloom's domestic imagination beyond the traditional Utopian framework, and argue that, rather than simply Utopian or consumerist, Bloom's domestic imagination is, in fact, both an adoption of the garden-suburb movement and a critical reflection of the Anglo-Irish Revivalist politics.

Inspired by H. G. Wells's comment on Joyce's "cloacal obsession"[19], the second chapter in this first section calls for attention to the cultural politics of waste and consumption, and contemplates the reconciliation of civilization, politics, and history through the languages of disposables. My consideration of waste matters opens with "Proteus," in which Stephen's walk on the polluted beach in Sandymount Strand "remembers" the history and the philosophical question of life and its transformation in the life cycles (*U* 3.147–67), and proceeds with Bloom in "Calypso" as his thoughts linger upon the liquid waste as garden fertilizer (*U* 4.475–84), as well as his eco-political reflections on waste, environment and resources in "Ithaca" (*U* 17.1698-1708). This leads to the political-environmental problem of the city's sewage system, introduced by the emergence of the civil engineer Tom Rochford (*U* 8.989), Professor MacHugh's talk on sewer and civilisation (*U* 7.487–98), and the lengthy segments on the provision and the flow of water in Dublin in "Ithaca" (*U* 17.163–228). How do these references to waste, sewage, and disposables throughout *Ulysses* reflect the anxiety of modernity about civilization? As Pierre Guyotat comments: "when I admit to preferring public shit to private shit, I am simply opposing the totalitarianism of shit to the totalitarianism of the State"[20]. Do these "shit matters," in turn, reinforce the modern totalitarian structure of politics in turn-of-the-century Ireland, or is Joyce proposing a "shitty" way of reconciling politics and history through an eco-system, sewage, and disposables? As these questions are being raised throughout the discussions in this chapter, I also look intertextually at Joyce's use of the "Agendath Netaim" throwaway and "plumbing consciousness," or "sewer of consciousness"[21], in constructing an alternative narrative of the modernizing city of colonial Dublin. I propose, from incorporating econo-political imaginations of wasteland plantation inspired by *Zionism and the Jewish Culture* to his referential use of the colonial politics of the modernisation of the sewage system in fin-de-siècle Dublin, Joyce writes of waste in a way that challenges the binary civilisation/waste distinction, redeems the narrative voice of the overlooked, and renders beautifully audible the languages of an ecosystem of his own making.

The second part of this book, entitled "Nature, Nation, and Eco-system," moves to examine the diverse ways in which nature and nation are (re)imagined exemplarily in Joyce's writing of the tree wedding procession and the marketplace. The section begins with a chapter on the gigantic tree wedding catalogue in "Cyclops." Focusing on the Tree Wedding scene in the episode of "Cyclops" (*U* 12.1258–95) in *Ulysses*, this chapter delves into Ireland's troubled history of forestry and the propaganda of afforestation at the turn of the twentieth century. It attempts a close reading of the wedding procession of "the chevalier John Wyse de Neaulan, grand high chief ranger of the Irish National Forests, with Miss Fir Conifer of Pine Valley" (*U* 12.1267–69), and proposes a critical discussion of the intertextual technique, the implicit cultural criticism, as well as Joyce's view of contemporary environmental politics in Ireland. While the promotion of or opposition to afforestation in Ireland reflects the historical contexts of the timber trade, the land problem, and the British-Irish struggle, these christened trees, I propose, may also recall the ancient Celtic social order of Brehon Law, in which species of trees were stratified into four classes, replicating the social order according to their economic importance.

Upon close examination, Giambattista Vico's historiographical study of the origin of *lucus* in *The New Science* sheds further light on the passage's specific location within the text. Situated between "Europe has its eyes on you" (*U* 12.1265) and "And our eyes are on Europe" (*U* 12.1296), the catalogue not only textually becomes the Cyclopean eye in Joyce's contemporary Irish context, but more significantly one with the "beam in the eye." In *Forests: The Shadow of Civilization*, Robert Pogue Harrison indicates the assimilation of Vulcan and Cyclops by Vico in *The New Science*, and he further notes how for Vico the Homeric concept of the Cyclopean eye, in fact, comes from the heroic phrase "every giant had his *locus*," the term which actually contains a dual meaning of "clearing" and "eye." I hence refer, in the second part of this chapter, to Vico's discussions on the myth of Vulcan and Cyclops, as well as forest clearing and patriarchy, to reconsider the Cyclopean catalogue's reflections on colonial/capitalist exploitation of the environment, deforestation,[22] and Joyce's rewriting of the nationalist movement in Ireland at the turn of the century. Curiously enough, in the episode's first published version, *The Little Review* installment issued in 1919, there was barely a tree wedding "catalogue." Apart from the wedding couple, only the maids of honor, Miss Larch Conifer, and Miss Spruce Conifer were present. Drawing on genetic evidence, this chapter attempts to decode Joyce's augmentation with regard to the political background of the afforestation agenda in late 19th-century Ireland alongside the passage's correspondence to ancient Celtic laws of patriarchy. Bearing in mind Joyce's augmenting process, in this chapter I will examine the tree wedding catalogue in light of Ireland's environmental politics and its political environment, and attempt to situate the tree list not just culturally and historically but also environmentally, politically, and historiographically.

The last chapter of the second section takes us back to the question of nature, politics, and consumption at the beginning of "Cyclops" where coming into view with Barney Kiernan's pub is the picturesque catalogue of an idyllic landscape. Rewritten by Joyce according to James Clarence Mangan's poetic translation of "Prince Aldfrid's Itinerary through Ireland," this idyllic catalogue is a rewriting of the modern scene of the Dublin Corporation Fruit, Vegetable and Fish Market in pastoral language. In this chapter, I consider the complexity of this passage, with its multiple layers of narrative schemes and retelling, and closely look at how inter-textual reading untangles the rich cultural meanings of this pastoral catalogue. Joyce's use of the Manganesque allusion, apart from signaling and satirizing Anglo-Irish Revivalists' "pseudo-bardic" style[23], further complicates his writing of the politics of the pastoral in a marketplace context. Drawing intertextual references from the "Aeolus" chapter as well as other pastoral traditions, the juxtaposed reading of these passages brings a new light to my interpretation of nationalistic rhetoric, the idyllic, and the culture of commerce. This then develops into the last part of this chapter, which focuses on the social geography of the Dublin Corporation market. Inspired by genetic criticism, journalistic resources, and other relevant studies on cultural geography, I propose a social reading of the catalogue in the context of the emergent culture of commerce in turn-of-the-century Dublin. Not only does the fish list appear like a catalogue of commodities for trade in the market news, the Cyclopean "shining palace [with] crystal glittering roof" turns out to be a mock-heroic version of the Dublin Corporation market. Contemplating further on the politics of colonial space of commerce, the chapter concludes with the Bakhtinian notion of social languages and argues that, with his writing of the pastoral in *Ulysses*, Joyce has awakened diverse levels of critical considerations for the modern relationship between languages and eco-politics.

As Ken Hiltner points out in the recently published collection *Ecocriticism: The Essential Reader* (2015), "Because of the [emerging] environmental justice movement, ecocriticism greatly benefited from the work of literary critics exploring issues like gender, class, race, and colonialism. Ecocritics are now returning (and will very likely in the future continue to return) the favor by showing how an environmental approach can enrich critical work in the fields, such a[s] colonial studies, from which environmental justice borrowed. In this sense, ecocriticism will, like the methodological approaches that preceded it, both remain a discrete field of literary study and inform other approaches"[24]. My goal for this current study is not so much to fill in the gap between Joyce scholarship and ecocritical studies but to call for attention to the ecologies of Joyce's major text and its social and political contexts. I have, however, demonstrated through the pages to follow that Joyce is nonetheless a writer with the environment in mind, and that the imagination of nature in *Ulysses* is inseparable from that of the emergent nation of fin-de-siècle Ireland. In this way, this project suggests an eco-political reading of *Ulysses* and hopes to, in due time, ignite future discussion on Joyce's environmental politics.

Anne Fogarty's foreword to the *Eco-Joyce* collection beautifully sums up the prospect of an eco-political reading of Joyce's work:

> Joyce's works, in their resistance to myths of a harmonising, recuperative nature, their emphasis on biopolitics, their blurring of the human and the non-human, their depiction of landscapes and urban settings in particular as sites of conflict and loss and their quest to voice the post-human, invent global idiolects, and reach beyond the limiting confines of the subject, invite ecocritical inquiry even as they also urge us to rethink many of the contentious precepts governing this critical field. Above all, they alert us to problematic mystifications of nature and organic life still dominant in Western culture and attendant concepts of the aesthetic, and to the need to expand notions of environmentalism to include the urban and the abject.[25]

What does an eco-political reading of *Ulysses* look like? It looks like a middle-class Jewish Irishman's gardening daydream, which expands into an eco-socialist urban planning scheme. It looks like a waste on Sandymount Strand, and sewage in the streets of "dear dirty Dublin," revealing the "shitty matter" of the modern totalitarian structure of politics in fin-de-siècle Ireland. It looks like a journalistic report on the gigantic arboreal clan in attendance at a family wedding, where natural history, nationalistic expressions, Brehon tradition, and Viconian historiography merge. It looks like a pastoral parody of romantic Revivalist nature writings, which blends itself with the languages of consumerism and the colonial spatiality of the Dublin Corporation Fruit, Vegetable and Fish Market. What does an eco-political reading of *Ulysses* look like? It looks almost like Joyce's own ecosystem of languages of nature, nation, and consumption, where "[t]he war is in words and the wood is the world" (*FW* 98.34–5).

Notes

1. William Wordsworth, "I Wandered Lonely as a Cloud" (1807), in *Penguin Book of Romantic Poetry*, ed. by Jonathan Wordsworth, p. 385.
2. From John Keats, "On the Grasshopper and the Cricket."
3. Michael J. McDowell, "The Bakhtinian Road to Ecological Insight," in *The Ecocriticism Reader: Landmarks in Literary Ecology*, eds. by Cheryll Glotfelty and Harold Fromm (Athens: University of Georgia Press, 1996), p. 375.
4. Enda Duffy, "Disappearing Dublin: *Ulysses*, Postcoloniality, and the Politics of Space" in *Semicolonial Joyce*, ed. by Derek Attridge and Marjorie Howes (Cambridge: Cambridge University Press, 2000), p. 46.
5. As also illustrated in Yeats' short play *The Land of Heart's Desire*. Marjorie Howe notes, "Dublin's incomplete and uneven colonial modernity had a counterpart in what we might call the perverse modernity of the Irish countryside. Rural villages in post-Famine Ireland were not modern anonymous collectivities. But they were also not the kind of totalizable, knowable, face-to-face community that some scholars associate with precapitalist forms of social life. The main reasons for this were Ireland's uniquely high rate of emigration, driven in part by agricultural modernization, mostly (during this period) to the United States, and the specific cultural meanings attached to emigration in Irish culture" (64).
6. For detailed information, Part II of A New History of Ireland: Vol. 6, Ireland under the Union 1870-1921 (Oxford: Oxford University Press, 1989), entitled "The Land War and the Politics of Distress, 1877-82," written by R. V. Comerford, offers an extensive study of the politics of land reform during this period in Irish history. Other useful resources included F. S. L. Lyons's Ireland Since the Famine (Harper Collins, 2009), Robert Kee's The Green Flag: A History of Irish Nationalism (UK: Penguin, 2000) and Ireland: A History (Abacus, 2003), Roy F. Foster's Modern Ireland, 1600-1972 (Penguin, 1990), and Thomas Bartlett's Ireland: A History (Cambridge University Press, 2010).
7. Terry Eagleton, *Heathcliff and the Great Hunger: Studies in Irish Culture* (Verso, 1995), pp. 8–9.
8. Ibid., p. 6.
9. John Wilson Foster, "Preface" in *Nature in Ireland: A Scientific and Cultural History* (Dublin: Lilliput Press, 1997), p. ix.
10. Cheryll Glotfelty and Harold Fromm, "Introduction" in *The Ecocriticism Reader: Landmarks in Literary Ecology*, eds. by Cheryll Glotfelty and Harold Fromm (University of Georgia Press, 1996), p. xix.

11. William Rueckert, "Literature and Ecology," in *The Ecocriticism Reader: Landmarks in Literary Ecology*, ed. by Cheryl Glotfelty and Harold Fromm (Athens: University of Georgia Press, 1996), p. 121.
12. Lawrence Buell, *The Future of Environmental Criticism: Environmental Crisis and Literary Imagination* (Blackwell, 2005), pp. 22-23.
13. Ken Hiltner, ed., *Ecocriticism: The Essential Reader* (London: Rutledge, 2015), pp. 132-33.
14. Gerry Smyth, "'Shite and Sheep': An Ecocritical Perspective on Two Recent Irish Novels" in special issue of *Contemporary Irish Fiction, Irish Universeity Review* 30:1 (2000), p. 164.
15. Sean Latham, "Twenty-first-century critical contexts," in *James Joyce in Context*, ed. by John McCourt (Cambridge: Cambridge UP, 2009), p. 156.
16. Rob Ware, "*Eco-Joyce*: The Environmental Imagination of James Joyce," Book Review in *Interdisciplinary Study of Literature and Environment* (2014) 21.4, p. 932.
17. William Kupinse, "Scrupulous Greenness: Eco-Joyce: The Environmental Imagination of James Joyce," Book Review in *BREAC: A Digital Journal of Irish Studies* (University of Notre Dame). <http://breac.nd.edu/articles/56844-scrupulous-greenness/>
18. Martin Ryle, "Eco-Joyce: The Environmental Imagination of James Joyce," Book Review in *Textual Practice* (2014) 28.6, p. 1156.
19. H. G. Wells, "James Joyce," in *New Republic 10* (10 March 1917), p. 159.
20. Qtd. in Dominique Laporte, *History of Shit*, trans. by Nadia Benabid and Rodolphe el-Khoury (Cambridge: MIT Press, 2000), p. 66.
21. Michael Rubenstein, *Public Works: Infrastructure, Irish Modernism, and the Postcolonial* (Uni. of Notre Dame Press, 2010), p. 70.
22. See *Capital: A Critique of Political Economy, Vol. II*: "The development of culture and of industry in general has ever evinced itself in such energetic destruction of forests that everything done by it conversely for their preservation and restoration appears infinitesimal." (Marx 1956: 248)
23. Andrew Gibson, *Joyce's Revenge: History, Politics, and Aesthetics in* Ulysses (Oxford: Oxford UP, 2002), p. 107.
24. Hiltner, p. 133.
25. Anne Fogarty, "Foreword," *Eco-Joyce: The Environmental Imagination of James Joyce*, eds. by Robert Brazeau and Derek Gladwin. (Cork: Cork UP, 2014), p. xviii.

PART I NATURE, LAND, CONSUMPTION

1 GARDENS
Irish Land Reform, Garden Cities and the Politics of Gardening[1]

> *Could Bloom of 7 Eccles Street foresee Bloom of Flowerville?*
>
> U 17.1581

The episode in *Ulysses* often referred to as "Ithaca" is one of homecoming, of receding into domesticity but also of science and information, and of the impersonal catechism. It is, when it comes to pages, the second longest one in *Ulysses*, second only to "Circe." Vincent Cheng describes it as "an episode that refuses to 'imagine' false identities, revealing instead a plethora of specific facts and objective details which are thus cleared of the suspicion that they might be either slanted by an individual stream of consciousness (in subjective indirect monologue), or exaggerated through stylistic parody or fantasy"[2]. Hence, receding from the world of movements and languages, "Ithaca" brings us toward (or backward to) the world of data, facts and omniscient information. It is the episode of descriptive scientific language: the narrative voice, or whoever puts questions and answers, sounds detached and impersonal.

Nevertheless, to be "impersonal" is not necessarily to be objective, while the catechizing process can be more information-bombarding than question-clarifying. In his discussions on the question of history in Joyce's work, James Fairhall brings in the fallacy of history as reality and introduces cross-questioning as an academic practice of history. He quotes from English philosopher-historian Robin Collingwood that: "The questions we ask about the past are determined by our own particular present, and the resulting answers — while never yielding full, absolute knowledge — can illustrate the past in terms of the present and vice versa"[3]. In his view, history is constantly being reconstructed through questioning and answering, and by "cross-questioning," certain unconsciously withheld information can be extracted. Joyce's "Ithaca," with its question-and-answer — and sometimes even cross-questioning — format, is in this sense a rearrangement of the fragments throughout the day of June 16, 1904, and beyond. It is the rearrangements — like Bloom's rearrangements of his pockets and receipts — of the significant (and insignificant) incidents and thoughts of the day, of impersonal knowledge and meticulous descriptions, of thoughts on race, religion, and history. The question-and-answer format is meant to disentangle the chaotic narratives of previous episodes and pull together a microcosmic narrative fabric of history.

Whereas Q&A and cross-examination practices help historians procure relevant information to mend historical gaps, the bombarding information drawn from the Q&As in "Ithaca" can be loquacious and overwhelming. Questions are usually given lengthy answers consisting of detailed, sometimes verbose, lists. Narrative flow from question to question remains so fluent and spontaneous that no intervention can possibly be made. Declan Kiberd tells us that "both the catechism and the science textbook had the same disadvantage: they ask a question not out of genuine uncertainty but only because the answer was already known"[4]. This, as Kiberd claims, is a form of "*interrogation*"[5], with which "the answers are already known, and the 'right' answer must be given, even if that is not what the interrogated person believes." Readers are subsequently made "mute" by the overwhelming blocks and terse interrogation, since they are not offered a chance to halt or think or question. Accordingly, "Ithaca" is an episode of a single narrator, yet he is a dogmatic one; it is a chapter that aims at the uncovering of information in order to access truth, yet it is also "a savage commentary on the overload of information in our modern world, information which oppresses more often than it illuminates"[6].

However, contrary to the episode's seemingly deliberate neglect of its readership in its impersonal catechistic interrogations, the writer Joyce was not only fully aware of his readers but of the fact that on a certain level, he wrote for an anticipated contemporary readership, especially that of the Irish Literary Revival. Joycean critics including Brian Caraher, John Nash, and Clare Hutton have explored the way Joyce, particularly in "Scylla and Charybdis," engaged in an interactive readerly response to Revivalist reader-writers including John Eglinton, George Russell (AE), and Oliver St. John Gogarty[7]. While the National Library episode is drawn from a private encounter between Joyce and Eglinton in the National Library of Ireland in 1904, it was significantly the moment in which Eglinton refused to publish Joyce's early manuscript of *A Portrait* in *Dana*, a journal of which Eglinton was the editor. The Library scene, Nash argues, can be read more than simply as "a restaging of that private moment" between Joyce and Eglinton in the National Library. It should also, as he persuasively points out, "be recalled that there is also a broader cultural and literary dynamic at work here"[8]. Whereas Nash, Hutton, and Joseph Kelly have brilliantly detailed at length the cultural politics and the question of readership in the Library scene, another editor present in the same setting, AE (George Russell) of *The Irish Homestead*, nevertheless remains unattended to in critics' discussions. The term the "pig's paper" (*U* 9.321), referring to the agricultural magazine mentioned above, does promptly flash through Stephen's mind during the conversation, nonetheless, the socio-political dimensions of this agricultural periodical evoked by its simultaneous presence in the National Library have been overlooked in Joycean scholarship thus far.

This chapter will focus on the episode of "Ithaca" for a close reading of Bloom's imaginary dwelling, "Bloom's Cottage" (*U* 17.1580), and his renewed identity as "Bloom of Flowerville" (*U* 17.1581), in view of the inherent reciprocity between agricultural co-operative propaganda and the politics of the Cultural Revival at the time. As I will elaborate later, Bloom's domestic imagination is part of yet not exclusively about the consumerist response to popular culture and advertisements in the modernizing Irish fin-de-siècle society. His consumerist desire in fact, I argue, responds to the propaganda set by the *Irish Homestead* and the I.A.O.S. (Irish Agricultural Organisation Society), which represents, significantly, along with the Gaelic League and the Irish Literary Revival, an essential cultural-economic dimension of the Co-operative movement to promote technical modernization, the Irish language, and cultural identity in Ireland at the turn of the twentieth century.

I.

Much of the information presented in "Ithaca" has to do with Bloom's (as well as Joyce's) attentive attitude toward advertisement and commodity culture, which, in Fairhall's word, is synonymous with "capitalism" – "the advance guard of the global capitalist economy now reshaping not only Irish lives but everyone's life"[9]. Critics have pointed out how Bloom's material desire for commodities forges his fantasy of

> a thatched bungalowshaped 2 storey dwellinghouse of southerly aspect, surmounted by vane and lightning conductor, [...] halldoor, olive green, with smart carriage finish and neat doorbrasses, stucco front with gilt tracery at eaves and gable, rising, if possible, upon a gentle eminence with agreeable prospect from balcony with stone pillar parapet over unoccupied and unoccupyable pastures and standing in 5 or 6 acres of its own ground [...] (*U* 17.1504-11)

And so on. In their view, Bloom's dream of a comfortable country life in an agreeable house is a reflection of his domestic desire. Fairhall, for one, describes the language here as that "of desire and imagined identity that characterizes the advertising copy and the articles of magazines devoted to elegant living"[10]. It is worth noting that Bloom chooses "[n]ot to inherit [...] gravelkind of borough English, or possess [...] an extensive demesne of sufficient number of acres, [...]

nor [...] a terracehouse or semidetached villa, [...] but to purchase by private treaty in fee [...]" (*U* 17.1499–1504) a suburban house. In this sense, he is not only dreaming of an ideal property as an object to possess, but also proclaiming his ability to engage in a consumer culture.

According to Jennifer Wicke, "[e]very object is also a relation, implies a work of consumption, a transforming recontextualization of the sort that goes on even with the more mundane goods of actual purchase: in Bloom Cottage, Saint Leopold's, Flowerville, [...], a whole range of philosophical and leisure activities are also suddenly possible"[11]. By this, she refers to the following series of questions and answers, ones that further detail Bloom's architectural plans and mechanical appliances, "Bloom of Flowerville," as well as the lists of "intellectual pursuits" and "lighter recreations" (*U* 17.1581, 1588, 1592). Ellen Carol Jones further elaborates on Wicke's point about consumption, reminding us that it is a "lifestyle," rather than material commodities themselves, that Bloom constructs in his ever-expanding Flowerville fantasy[12]. Such lifestyle, in the following Q&As, would make him "a gentleman of field produce and live stock," and obtain "ascending powers of hierarchical order, that of gardener, groundsman, cultivator, breeder, and at the zenith of his career, resident magistrate or justice of the peace with a family crest and coat of arms and appropriate classical motto [...], duly recorded in the court directory [...], and mentioned in court and fashionable intelligence" (*U* 17.1603, 1608–14). As his social status ascends and his political "capacity" increases, Bloom furthermore outlines for himself a political "course of action" (*U* 17.1616), which is

> A course that lay between undue clemency and excessive rigour: the dispensation in a heterogeneous society of arbitrary classes, incessantly rearranged in terms of greater and lesser social inequality, of unbiassed homogeneous indisputable justice, tempered with mitigants of the widest possible latitude but extractable to the uttermost farthing with confiscation of estate, real and personal, to the crown. Loyal to the highest constituted power in the land, actuated by an innate love of rectitude, his aim would be the strict maintenance of public order, [...] the upholding of the letter of the law [...] against all traversers in covin and trespassers acting in contravention of bylaws and regulations, [...] all orotund instigators of international persecution, all perpetuators of international animosities, all mental molestors of domestic conviviality, all recalcitrant violators of domestic connubiality. (*U* 17.1617–33)

The passage is Bloom's tactful political statement on social equality, as well as the redistribution of property. He advocates land reform, speaks for Home Rule, and assents to the enforcement of violent control when necessary. These claims, however, have been similarly voiced in "Circe," when Bloom declares his political ideals for his Utopian regime of a "new Bloomusalem":

BLOOM

I stand for the reform of municipal morals and the plain ten commandments. New worlds for old. Union of all, jew, moslem and gentile. Three acres and a cow for all children of nature. Saloon motor hearses. Compulsory manual labour for all. All parks open to the public day and night. Electric dishscrubbers. Tuberculosis, lunacy, war and mendicancy must now cease. General amnesty, weekly carnival with masked licence, bonuses for all, esperanto the universal language with universal brotherhood. No more patriotism of barspongers and dropsical impostors. Free money, free rent, free love and a free lay church in a free lay state. (*U* 15.1684–93)

Whereas Bloom's political statement in "Ithaca" is uttered in an objective and understated language, "Circe"'s phantasmagorical setting gives more rhetorical freedom to his lecture on politics. Such freedom not only grants him the independence of speech but also allows him to "play out in unrestricted imagination his ultimate utopian fantasies as an Irish Messiah and reformer"[13]. However, the surreality of this ambitious speech may not be as ridiculous as it seems: in fact, as Gifford reminds us, "three acres and a cow" was a phrase that became "the rallying cry for Irish land reform after its use by Jesse Collings," a member of the Parliament, "in a successful effort to force a measure of land reform on Lord Salisbury's conservative and reluctant government in 1886"[14]. That is to say, here Bloom is advocating "an equitable land reform program that redistributes Irish territory to the Irish"[15]. In Bloom's proposed policies in "Ithaca," he also calls for the "dispensation [...], incessantly rearranged in terms of [...] social inequality, of unbiased [...] justice, tempered with mitigants of the widest possible latitude but extractable to the uttermost farthing with confiscation of estate, real and personal, to the crown" (*U* 17.1618–22). With its circuitous language, the passage also proclaims the need for land reform in Ireland.

It is not coincidental that both passages of Bloom's prospective political policies involve land reform. As Joseph Lee signals: "Post-Famine Ireland had a land question. It had no peasant question"[16]. The Irish land question is based upon years of the country's economic reliance on agriculture and crop exports; such reliance turned thorny when the country was struck with a Great Famine and successive agricultural depressions. The problems of the Irish landlord-tenant system lie in unaffordable high rents (especially during depression years), strict Land Acts, and the tension between Irish tenants and British absentee landlords. On the other hand, the already questionable landlord-tenant system in Ireland became more problematic during the depressions, especially that of 1879-82, when the unadjusted rent exceeded tenants' ability to submit payments during those difficult years[17]. Michael Turner comments on the way agricultural economic history influences the political history of Ireland by noting that it was the 1879-82 depression and the associated Land War which "exposed the tensions at the opposite poles of the social and economic ladder, [...] led to concerted political moves towards Irish home rule," and eventually heralded the subsequent move to independence[18]. The land problem in Ireland, therefore, is not only part of the national paralysis that leads to political reforms; it is among the foundations of a national economy, and it lies at the heart of Irish nationalism and the independence movement.

Joyce sets *Ulysses* in the year of 1904, only a year after Wyndham's Land Act of 1903. This time frame bears a significant meaning not only regarding the rising political tensions as the aftereffects of the Home Rule Bills and the Phoenix Park Murders in 1882, but also those in relation to the Land League, nationalism and, in the following years, the rise of property ownership. The shocking incident of the Phoenix Park Murders, as a turning point in the Irish-British political relationship, came as a response to the Kilmainham Treaty, signed between Gladstone and Parnell as an extension of the 1881 Second Land Bill[19]. Evidently, the Irish land problem is the socio-political driving force that propels the development of Irish nationalism and the independence movements of the country.

In "Ithaca," right after Bloom's political "course of action" (*U* 17.1616) and the claim of his own "innate love of rectitude" (*U* 17.1623), the narrator requests a proof that "he had loved rectitude from his early youth" (*U* 17.1634). Then comes a brief account of his religious and political development since adolescence; among the anecdotes, it is narrated how "In 1885 he had publicly expressed his adherence to the collective and national economic programmes advocated by James Fintan Lalor, John Fisher Murray, John Mitchel, J. F. X. O'Brien and others, the agrarian policy of Michael Davitt, the constitutional agitation of Charles Stewart Parnell [...], the program of peace, retrenchment and reform of William Ewart Gladstone [...]" (*U* 17. 1645–51). Whereas Bloom's support for Gladstone's "peace, retrenchment and reform" programme corresponds to his own proclaimed policy with "measure of reform or retrenchment [...]" (*U* 17. 1625), some of the political figures he claims to support closely relate to the Land League. The Irish writer James Fintan Lalor, for one thing, "vigorously advocated republicanism and a radical program of *land nationalization*"[20]. The initiating organizer of the Land League, Michael Davitt, on the other hand, had a land reform programme which "advocated the use of public funds to achieve peasant ownership of the land"[21]. It is interesting that Bloom doesn't simply agree with Davitt's land policies: in the "Eumaeus" episode, as a "backtothelander" (*U* 16.1593) he furthers Davitt's proposal "by advocating an agrarian socialism in which all men would contribute by sharing agrarian labor"[22]. Such propaganda, familiar as it sounds, corresponds to his political statement of land distribution and shared labor, announced phantasmagorically in "Circe," and later paraphrased, in "Ithaca."

II.

> *It is in the cottages and farmers' houses that the nation is born.*
>
> — George Russell (AE), *the Irish Homestead*

Despite Joyce's (and Bloom's) attentiveness to the Land League and land reforms, and despite Irish nationalism's disturbing relationship with the land problem, Irish rurality curiously remains outside the major narrative frame in *Ulysses*. James Fairhall observes that "since the closest model for Joyce's collection [*Dubliners*] was George Moore's *The Untilled Field* (1903), set largely in rural Ireland"[23], there must be an intentional omission of the countryside. In his opinion, "[Joyce's] uneasy relationship with Ireland, especially rural Ireland, is [...] presaged by his first publication, which in the context of the *Homestead* (the "pigs' paper" [*U* 9.321]) enters into uneasy dialogue with a world of cream separators and butterflies among thistles which 'Stephen Dedalus' clearly has judged and found wanting"[24]. However, if Stephen Dedalus indeed finds the journal wanting, how would Leopold Bloom, a man of science, politics, and advertisement, find the *Irish Homestead*?

First released in 1895, the *Irish Homestead* was the journal of the Irish Agricultural Organisation Society (IAOS), the co-operative agricultural society founded in 1894 by Sir Horace Plunkett to "better the material circumstances of the emerging class of small farmers"[25]. Around 1904, at a time of significant land transfer as a result of the Land Acts, the IAOS played a vital role in educating the new incumbents in modern farming, while "[w]eek-in week-out, the *Irish Homestead* urged a program of social reform that constantly pitted a desirable middle-class propriety against the uncouthness of certain traditional practices"[26]. F.S.L Lyons notes "it was essential to Plunkett's concept of co-operation that while the IAOS should be propagandist in the agricultural sense, it should be politically neutral"[27]. However, it is important to keep in mind that whereas political neutrality was the purpose of the organization, its practice inevitably involves the co-operative politics on the basis of which the society was first formed. P. J. Mathews argues that Plunkett's agricultural society was part of the "co-operative movement" which, promoting a republican self-help ethos against the constitutional nationalism of the Irish Parliamentary Party, led to the establishment of the Abbey Theatre, the Gaelic League, the IAOS, and the

National University of Ireland. With the defeat of Gladstone's second Home Rule Bill in 1893 and the passing of the Local Government Act in 1898, post-Parnell Ireland was under the Conservative threat to "kill Home Rule with kindness"[28]. Against this background, these later co-operativists realized that "little could be achieved at Westminster" and as a result came up with "the strategy of working for a form of *de facto* home rule despite its unattainability *de jure*"[29]. Their plan was "to mobilize and apply the latent national intelligence of the country to the practical needs of Ireland, a strategy conveniently encapsulated in the term 'self-help'"[30]. Not surprisingly, this nationalistic self-help ethos was also in tune with the Revivalists' call for the cultural revival of Ireland. Many well-known Revivalists including Lady Gregory, W. B. Yeats, J. M. Synge, Douglas Hyde, and, above all, George Russell (AE), were among the activists involved in this co-operative movement. Indeed, as early as in the 1898 article "Ireland, Real and Ideal," Lady Gregory identified the revival's dual aim in material-alongside-cultural productions with the partnership of Sancho Panza and Don Quixote in *Don Quixote*[31]. A decade later during the early years of the twentieth century, the Gaelic League committed itself to "support for the industrial revival (setting up an industrial committee in 1902), Irish temperance, technical education, *agricultural co-operation* and support for Irish as a qualification for public bodies in the Gaeltacht"[32]. The agricultural co-operation promoted by the IAOS, hence, is contemporarily regarded not as a separate economic scheme but one of the contributing forces to the larger cultural nationalist movement of the Celtic revival.

P.J. Mathews regards the period between 1899 and 1905 as a "progressive [one] that witnessed the co-operation of the self-help revivalists to encourage local modes of material and cultural development"[33]. He refuses to read the tension within early-twentieth-century Irish nationalism in the light of a cultural and political dichotomy[34], and instead proposes to approach it "as a battle between a newly emerging self-help consensus and old-style parliamentary politics"[35]. Despite his elaborate analysis of the opposition of co-operative politics to constitutional nationalism during the Irish Revival, Mathews fails to specify the dynamics of Anglo-Irish dominance in the co-operative movement, especially since it was also an unsettling epoch which witnessed the Revival's internal conflict between the Anglo-Irish Ascendancy and Catholic Ireland. Critics including Len Platt, Andrew Gibson, and Clare Hutton have detailed Joyce's intricate workings of Anglo-Irish politics and the Revivalist institutions in his composition of *Ulysses*[36]: whereas Platt calls for consideration for the problematics of materialism and revivalist nationalism in "Ithaca," Gibson argues that the scientific language in this very episode exhibits the politics of the technical culture of the sciences dominated by Anglo-Irish Protestant revivalists. Despite their respective focuses, both Platt and Gibson attend to an (Ithacan) scientific discourse that is technically objective but unevenly distributed among classes and historically imperial in practice.

For Len Platt, civilization, as "the precise and unequivocal locus of *Ulysses* 17," is one that "thinks of itself as being the product of a 'natural' evolutionary process, but which is emphatically middle class"[37]. He deems Joyce's depiction of his Irish homeland as a product of a deeply conventional, rationalist and materialist "middlecrass"[38]. Significantly, it is also this rising Catholic middle class that, according to Leeann Lane, troubles the Anglo-Irish co-operative promoters in their respective attitudes toward modernizing Irish society during the cultural revival. Fin-de-siècle Irish society was at the time undergoing industrial modernization while it observed the rise of the bourgeoisie as well as the transformation of social structures. Indeed, as John Hutchinson also illuminates, at the heart of the late-nineteenth-century cultural revival is the status anxiety of privileged Anglo-Irish Protestants. Confronted with the emergent urban Catholic bourgeoisie at a time when the compromising British government, after the Act of Union, was gradually releasing its authority over Ireland, the Protestant Ascendancy eagerly sought to maintain its elite position. The co-operative movement, in Lane's words, is in an essential regard "part of the larger movement of cultural revival initiated by the Protestant intelligentsia at the end of the nineteenth century as a response to their increasing identity crisis"[39]. Complementing P. J. Mathews' view of the IAOS and the *Irish Homestead* as part of the larger co-operative movement that responds to paralysing constitutional nationalism, Lane sees the agricultural co-operative magazine of the *Irish Homestead* as revivalist propaganda to "halt the full-scale implications of the growth of democracy and the rise of the new middle class in Ireland" by providing "an alternative leadership role for the Anglo-Irish as a newly reconstituted cultural aristocracy"[40].

Significantly, Bloom's imagined identity as "Bloom of Flowerville" (*U* 17.1581) details a progression in a hierarchical society not dissimilar to the Anglo-Irish co-operative blueprint. After a lengthy catalogue listing his material ambitions, Bloom envisions himself involved in "ascending powers of hierarchical order," from "that of gardener, groundsman, cultivator, breeder, and at the zenith of his career, resident magistrate or justice of the peace with a family crest and coat of arms and appropriate classical motto ..." (*U* 17.1608–11), and so on. This hopeful career outline that is ultimately emblematized with a family crest, coat of arms, a motto in Latin (*Semper paratus*), and respectable upper-class attentions, describes his speculative engagement in an aristocratic tradition of Ascendency lineage, one that, in Joyce's time, was exclusively Anglo-Irish.

It is indeed no coincidence that Bloom's auspicious "hierarchical ascendency" is conditioned by his ideal alternative career as a "gentleman farmer of field produce and live stock" (*U* 17.1603). Such an elevated image of a gentleman farmer, significantly concurrent with Plunkett's *noblesse oblige* motto, is one that the Anglo-Irish co-operativists attempted to publicise with their propaganda in the *Irish Homestead*. In an article entitled "The Irish Cottage," published in the weekly co-operative periodical on April 29, 1899, AE writes: "There is no more ideal life than the farmer's, no life which contains more elements of joy, mystery, and beauty"[41]. He claims that compared to the "insectiferous fakir and his kind" who "scorn the earth under their feet," the man is superior who "takes his patch of soil and labours on it until his world becomes as beautiful as other's dreams"[42].

AE's depiction of a romanticised farming class is, of course, no invention of his own. From the eighteenth century onwards, cultural nationalists in Ireland had been committed to a romanticised image of Irish farming; the leading linchpin of the early-nineteenth-century Celtic revival, George Petrie, is a case in point. Under the influence of Wordsworth's poetry, Petrie "found immanent in the unspoiled Irish landscape, embedded with the monuments of successive generations and giving colour and life to the remote rural communities working its soil, a unique moral vision that gave a special power and direction to the historical existence of the Irish people"[43]. Through the *Dublin Penny Journal* and the *Irish Penny Journal* which Petrie was involved with[44], Petrie and his fellow journalists intentionally presented to the Irish public an image of the Irish farmer as "a man of the inner man," who "throughout his history displayed *the natural strength, intelligence and tenderness that belonged to man in his simpler rural state*"[45]. Romanticised farmer figures like this also found favor with the fin-de-siècle co-operativists as the latter alertly managed to preserve the Anglo-Irish Ascendancy's previously privileged position in economics, politics, and education against the threats posed, after Catholic emancipation and successive Land Acts, by the "rampant double-chinned vulgarity"[46] of the rising Catholic middle class. By aiming to regenerate morally the small-scale farmers in post-Land-Act Ireland, these agricultural co-operativists aimed to model a new class of the "noble peasant of the cultural revival" among the Anglo-Irish, a class that is not only morally and economically "noble," but would become "cultural and intellectual leaders" who "preserve 'the influence and place of *their class* in Ireland' "[47]. As if emerging out of AE's propaganda of the noble farmer, "Bloom of Flowerville" is a demonstrative figure of the "lifestyle" that is not only Bloom's personal ambition but which is correspondent also to the ideal middle-class countryman of the Anglo-Irish Ascendancy, an image the *Irish Homestead* hoped to publicise.

Central to the cultural revivalists' propaganda during the late-nineteenth-century Celtic Revival is their antagonism toward the material culture of industrialized England. This they demonstrated specifically with an ardent call to boycott British products in support of Irish co-operative manufactures. Indeed, such antipathy toward Anglicised material culture also has its root in the defining revivalist texts produced in the earlier generation by Petrie, O'Curry, O'Donovan, Samuel Ferguson and, above all, Standish O'Grady. Since then, Irish nationalists – both cultural and political – have tended to regard British materialism as the "dark corrupting forces" that have "wrecked ... Ireland's empire of spirit"[48]. However, such an antithesis ran into a challenge posed by the industrial economy of Ireland when the nineteenth century was approaching its end. When materialism and capitalism emerged as the consequences of the technical progression and urbanization in industrial England, the Irish nationalists' call for enmity toward Britain-oriented capitalism and materialism appeared suspiciously disadvantageous in the comparatively regressive Irish economic conditions. To this end Yeats, who considers it was Ireland that "contained a vision of the 'new man' whereas Britain represented a passé materialism," reciprocated with a regeneration scheme for a "new Ireland" that would "retain the apparatus of modern technology, but harnessed to the heroic integrative vision of its rural past"[49]. Somewhat in the same vein as Yeats regarding the revivalist spirit and idyllic rurality, AE's solution to reconcile the demands of modernity, industry, and a nationalist rural ideal lies in his co-operative programme, and above all found its voice through *The Irish Homestead*.

Interestingly, like this econo-political dilemma of the Anglo-Irish revivalists at the turn of the twentieth century, it is the scientific language illustrating material culture that constitutes one of the many layers of the perplexing contemporary politics in "Ithaca". On the other hand, class distinctions complicated the socio-political dimensions of such politics, since scientific discourse remained, as Gibson and Whyte observe, a Protestant practice despite the founding of many scientific academic institutions throughout the nineteenth and early twentieth centuries. However, specifically from 1908 onwards, after the Gaelic League was "wracked by internal dissensions and secessions in trying to counter [its decline]," D. P. Moran and other nationalists in turn "veered towards a rising 'neo-traditionalist' reaction against Ireland's exposure to materialist values"[50]. It is clear, then, that the rise of applied science and consumerism as a result of the demands of material culture challenged revivalists' mysticism and heroic romanticism, as well as the nationalists' anti-British, anti-materialist campaign.

In "Ithaca," Bloom's meticulous gardening self-portrait appears

> In loose allwool garments with Harris tweed cap, price 8/6, and useful golden boots with elastic gussets and wateringcan, planting aligned young firtrees, syringing, pruning, staking, sowing hayseed, trundling a weedladen wheelbarrow without excessive fatigue at sunset amid the scent of newmown hay, ameliorating the soil, multiplying wisdom, achieving longevity. (*U* 17.1582–1587)

Notice, in Joyce's depiction of "Bloom of Flowerville," Bloom's imaginary possession of the allwool cap is fabricated with "Harris tweed" (*U* 17.1582), a handwoven fabric made with local wool by the islanders on the Outer Hebrides of Scotland. Among other delicate furnishings in his domestic household plan, a "handtufted Axminster carpet" (*U* 17.1526) is self-evident of Scottish manufacture. Of course, it was 1904, still four years to go before the revivalists' conscious opposition to British materialism, and Griffith's advocacy of the purchase of local Irish wool as advertised and outlined in "Sinn Féin Policy"[51] will not be published until more than a full year later. Surely, on the one hand, such anti-British/antimaterial sentiment didn't come out of the blue at once, and on the other, this propaganda wouldn't be unaccessible to Joyce in 1921 when he was composing the "Ithaca" episode. In fact, in support of Griffith's programme, Joyce himself even intended to purchase Irish gowns for Nora when the Nationalist advertisement first appeared in *Sinn Féin* in 1906[52]. In effect, in contrast to the contemporary Nationalist scheme, Bloom's consumerist choice of Scottish fabric products anticipates the Anglo-Irish ambivalence toward materialism and consumerism in contemporary popular culture.

With the *Irish Homestead* and the IAOS, AE's call for regeneration of rurality through agricultural technical development appeared, during this era of conflicting values, as a will to compromise diverse political forces. In *Co-operation and Nationality*, a pamphlet first published in 1912, AE sensationally claims: "It may be said we are hoping to substitute an agricultural ascendancy for this urban ascendancy"[53]. What he suggests is a "new order" for Ireland, one that incorporates advanced agricultural industry, progressive co-operative economy, and communal mutual-help. This idea of the communal co-operation of AE's, as Nicholas Allen signals, owes significantly to the Russian anarchist-communist Peter Kropotkin. "Kropotkin had after all prophesied in *Mutual Aid* that the growth of voluntary organisations was evidence of co-operative evolution," and, Allen writes, "Russell chose this moment of increased solidarity, of the potential union between co-operation and forms of advanced socialism, to introduce an argument on behalf of rural labour to the *Irish Homestead*"[54]. Like AE, Joyce was not unfamiliar with Kropotkin. Attentive to the emerging political thoughts and movements not only in Ireland but also in Continental Europe, Joyce was acquainted with contemporary socialist and anarchist discourses and had even been engaged in Italian Irredentist meetings during his stay in Trieste.[55] Although he later stopped attending these socialist meetings on a regular basis, his interest in anarchism remains, as Manganiello illustrates in *Joyce's Politics*, as manifested in his writings. Among other anarchist thinkers, Joyce shared AE's interest in Kropotkin's writings and his anarchist-communalist thoughts, and had the latter's *The Commune of Paris*, *The Conquest of Bread* (in Italian), *Fields, Factories and Workshops,* and *The Grand Revolution* (again in Italian) in his library. Manganiello specifies the anarchist-communist influence of Kropotkin on Joyce, one of the examples being Bloom's political proclamation as well as his communal ideal in "Circe," where Bloom claims to stand for

> ... the reform of municipal morals and the plain ten commandments. New worlds for old. Union of all, jew, moslem and gentile. Three acres and a cow for all children of nature. Saloon motor hearses. Compulsory manual labour for all. All parks open to the public day and night. Electric dishscrubbers. Tuberculosis, lunacy, war and mendicancy must now cease. General amnesty, weekly carnival with masked licence, bonuses for all, esperanto the universal language with universal brotherhood. No more patriotism of barspongers and dropsical impostors. Free money, free rent, free love and a free lay church in a free lay state. (*U* 15.1685-93)

This fantasmagorical proclamation of Bloom's, as I have previously pointed out, anticipates, in "Ithaca", the political ambitions to be carried out by "Bloom of Flowerville" at the "zenith of his career" (*U*17.1617–33). But it is also an explicit evocation of the socialist idea of communalism and, in particular, Kropotkin's anarchist-communalism. In *The Conquest of Bread*, Kropotkin outlines an ideal anarchist community in which "all belongs to all" and "all things [are] for all"[56], where bridges and roads all belong to public property, where museums, libraries and schools all provide free entry, and where "parks and gardens [are] open to all" citizens[57]. Bloom's "Bloomusalem" outlines a similar "scheme for social regeneration" (*U* 15.1702–3), rendering it not simply a socialist utopia, but more specifically an anarchist community.

As Bloom's scheme for Bloomusalem anticipates the "course of action" to be adopted at the zenith of his career outlined in "Ithaca," his hopeful increasing powers which would initiate with his alternative identity as "Bloom of Flowerville" (*U* 17.1581), the "gentleman farmer of field produce and live stock" (*U* 17.1603), indeed not only concretises AE's ideal farmer figure in the *Irish Homestead*, but furthermore incarnates the lifestyle of a community member from Kropotkin's prospective anarchist community. Manganiello keenly observes that the Flowerville passage "reveals a penchant for 'industry combined with agriculture and brain work with manual work' "[58], a phrase he borrowed from the subtitle of Kropotkin's *Fields, Factories and Workshops*. Interestingly and perhaps not coincidentally, Bloom devises, for his Flowerville grounds, "a glass summerhouse" (*U* 17.1552), "an orchid, kitchen garden and vinery" (*U* 17.1559) among his many horticultural pursuits, while in *Fields, Factories and Workshops,* Kropotkin proposes a "new system of horticulture" with "the market-garden under glass"[59]. The idea of the "kitchen garden" to be moved under glass shelter for a more efficient communal production of "market-gardening" is one of Kropotkin's major schemes for his communal economy. He exemplifies in this book and previously in the "Agriculture" section of *The Conquest of Bread* how the beneficial "hothouse culture" of market-gardening — "the *growing of fruit and vegetables in greenhouses*" — has "taken a great development" in the Channel Islands as well as in suburban London[60]. "In fact," he illustrates quoting from the *Encyclopaedia Britannica*, "along the railways which radiate from London in all directions the glass-houses have already become a familiar feature of the landscape"[61]. This reminds us of Bloom's vision for his prospective Flowerville estate to be located ideally "at a given point not less than 1 statute mile from the periphery of the metropolis, within a time limit of not more than 15 minutes from tram or train line" (*U* 17.1514–6), and with a glass-sheltered summerhouse among his outdoor properties to be adopted, his greenhouse would ideally be one of the prospective glass-houses in view if Kropotkin's anarchist-communalist plan were to be carried out in suburban Dublin.

Throughout *The Conquest of Bread* and *Fields, Factories and Workshops,* Kropotkin renounces capitalist exploitation by the middleman, and his exclusion of hierarchical capitalism in the hope of reaching a communal society based upon equality becomes the foundational aim of his anarchist-communist propaganda. His designated mutual-aid system, constructed upon co-operation among farmers without the interference of the mediator, must have inspired AE in his ardent promotion of a co-operative agricultural community. Joyce too adopts a position sympathetic to their opposition to these exploiting middlemen. Indeed, in *Ulysses,* he writes with the reprimand of gombeen men in "Cyclops," and Bloom, appearing in Barney Kiernan's pub earlier simply to meet with Martin Cunningham in order to settle funds for Dignam's widow, encapsulates the similar spirit of mutual aid in a co-operative community.

It might seem self-contradictory at first glance to consider Bloom's Flowerville alongside AE's and Kropotkin's proposals of communal co-operation, since the former's attentive illustrations of materialistic consumer culture contradict the latter's opposition to British materialism and capitalism. However, it is precisely this contradiction that complicates AE's own agricultural propaganda. While his co-operative movement aims to alleviate the Anglo-Irish anxiety about a prospective promise of "rural ascendency," AE's indebtedness to Kropotkin's anarcho-communism suggests an inherent anti-hierarchical consciousness and set of commercial practices in the actual co-operative scheme. Well informed about Kropotkin's anarcho-communist agenda and personally acquainted with AE and the *Irish Homestead*, Joyce must have been conscious of this contradiction, and he responds with a mocking portrayal of Bloom of Flowerville, an alternative identity to be mediated, scientifically excusing himself, to "[alleviate] fatigue and [produce] as a result sound repose and renovated vitality" (*U* 17.1757–8) before he "[retires] for the night" (*U* 17.1757).

Ellen Carol Jones indicates that with "the homologous dynamics of interpreting the form of dreams and that of commodities within the textual display of commodities in Bloom's dream of Flowerville," he "explores the unconscious desire articulated in the commodity form and in the dream-work — unconscious desire as traumatic unsymbolizable psychic and social division, as the Real — masked by the fantasy that structures social reality"[62]. Bloom's household fantasy, though not so "unconscious," does register the commodious material desire that "structures social reality" in the form of a utopian alternative[63]. More complicated than Žižek's assertion about the utopian substitution of an individual's unquenched commodious desire, Bloom's imagination emerges corporately with a socialist ideal, one that was inspired by Kropotkin's anarchist community and propagandized by AE's vision for a noble farming life by means of *The Irish Homestead*.

III.

> *What might be the name of this erigible or erected residence?*
> *Bloom Cottage. Saint Leopold's. Flowerville.*
>
> (*U* 17.1579–80)

In "Ithaca," Leopold Bloom's "ultimate ambition" (*U* 17.1497) started with the "erigible or erected residence" of Bloom Cottage, which is envisioned as "a thatched bungalowshaped 2 storey dwellinghouse …" with detailed exquisite exterior designs and interior decorations (*U* 17.1499–550). I have previously commented on the consumer culture and materialism with which Bloom's domestic consumerist dream engages, and many critics have also elaborated on the consumer politics of this particular passage. Danis Rose, among others, points out the possible practicality of Bloom's plan to purchase this suburban property. In his 1987 article on Bloom's wealth, he draws from Joyce's manuscripts and placards as well as the mortgage information provided by "The Irish Civil Service Building Society" (*U-G* 1582), the title of which organisation Joyce used in the earlier manuscript before he replaces it with the sarcastic "Industrious Foreign Acclimatised Nationalised Friendly Stateaided Building Society" (*U* 17.1658–9), as appeared in the chapter's published version. Rose's genetic approach not only confirms his argument on the practicality of Bloom's plan of property purchase but, to our benefit, supports my hypothesis on the politics of Irish architectural history to be contemplated in this particular context.

The following section delves into the politics of the (Anglo-)Irish Revival and its links with the Co-operative movement, the garden suburban planning movement, and relevant discourses of Irish cultural nationalism, to consider the country-versus-city complex in the turn-of-the-twentieth-century Ireland. Taking the economics and politics of modern Irish architectural history into consideration, I propose a socio-political interpretation of Leopold Bloom's imagined residence of Flowerville beyond the traditional Utopian reading, and argue that, rather than simply Utopian or consumerist, Bloom's domestic imagining is, in fact, both an adoption of the garden suburb scheme and a reflection of Anglo-Irish Revivalist politics, one that witnesses the awakening consciousness of the problematics of urbanisation.

In his brief genetic study on Bloom's imagined residence, Danis Rose particularly signals a curious line in one of the earlier fair copies concerning the purchase of the Bloom Cottage:

Was his ambition practically realisable?
Yes. (*U-G* 1582)

According to the financial details of the mortgage quotation provided by the Irish Civil Service Building Society (ICSBS), Bloom's house-purchasing scheme could be fundamentally realizable. However, it is worth noting that although the financial plan may be practicable, the property detailed in Bloom's dream is, in fact, a peculiar one. Vivien Igoe, the author of *James Joyce's Dublin Houses & Nora Barnacle's Galway*, once advised me that it is not common to have a thatched bungalow-shaped house with two stories. Murray Fraser also notes "in general [...] the typical medieval cottage in Ireland had been single-storey, usually detached, and built in timber frame and thatch"[64]. Although one of the two post-Ulster-Plantation English influences upon housing designs since the seventeenth century was the addition of a second storey, the historical custom of Ireland still "militated against a departure from the single-storey dwelling"[65]. As a result, the trend for two-storey houses never really emerged until the nineteenth century, not to mention the rare case of a "thatched bungalowshaped 2 storey" one (*U* 17.1504–5).

Fig. 1.1: **Foremen's cottages at "Kilkenny Garden Village" at Sheestown, Co. Kilkenny (1907), designed by William A. Scott.** The illustration originally appeared in the prestigious Irish architectural periodical, *Irish Builder and Engineer*, on November 30, 1907. Source: archive.org/details/irishbuilderengi4919unse

Considering the peculiarity of such an architectural design in the Irish context, it is especially of significance to connect Bloom's ideal "thatched bungalowshaped 2 storey dwellinghouse" (*U* 17.1504–5) with the thatched bungalow-shaped two-storey cottages which actually appear at the Kilkenny Garden Village in Sheestown, County Kilkenny, Ireland (see Fig 1.1). These industrial "green village" cottages were designed in 1907 by William A. Scott (1871–1921), one of the most renowned Irish architects of the time. Scott's design for the Kilkenny Garden Village was undertaken in response to the request by the enthusiastic Gaelic Leaguer and Revivalist, Captain Otway Cuffe, hence the Revivalist undertone of this building scheme. Indeed, Scott was known for his Neo-vernacular style and was especially closely related to the Arts and Crafts movement[66] as well as the Irish Renaissance, even to the extent of being regarded as the "architect by appointment to the Celtic Revival"[67]. His plan for the Kilkenny garden village exemplarily boasts his heritage from the Arts and Crafts spirit, adopting features of vernacular forms and local materials for construction.[68] From around 1917, he was even employed by W. B. Yeats to supervise the restoration of Yeats' residence of Thoor Ballylee (Ballylee Tower) in County Galway. Yeats' restoration plan for the tower, drawn from the influence of his early years residing in Bedford Park, a garden suburb on the western edge of London which "aimed at a rural 'Old English' effect, with trees and birds and pretty cottage gardens framing 'artistic' houses for people with aesthetic leanings and little money," translates a good deal of this garden village ideal into, in Sheehy's words, "a less suburban Irish idiom, with a bit of Irish history thrown in for good measure"[69]. Remembering his first experience of moving into Bedford Park, W. B. Yeats writes: "We went to live in a house like those we had seen in pictures, and even met people dressed like people in the story-books. The newness of everything, the empty houses where we played at Hide-and-seek, and the strangeness of it all, made us feel that we were living among toys"[70]. Such is the picture of an innocent, idyllic community after which Yeats laid out his blueprint for Thoor Ballylee. However, if Thoor Ballylee could be read as the idyllic ideal of the garden suburb transformed, the green village scheme in Kilkenny is, of course, another relevant model under the impact of the garden suburb aspiration. In his imaginable Flowerville residence, Bloom pictures himself "ameliorating the soil, multiplying wisdom, [and] achieving longevity" with a lifestyle which involves various simultaneous intellectual pursuits (U 17.1588–91) and other lighter recreations including "vespertinal perambulation or equestrian circumprogression with inspection of sterile landscape and contrastingly agreeable cottagers' fires of smoking peat turves (period of hibernation)" (U 17.1592–1602), etc. Such a relaxed idyllic image of suburban life interestingly concurs with the suburban garden experience in Bedford Park Garden Suburb as Yeats once favorably recounts. By no means coincidental in their similar idyllic ideal, the parallel between Yeats's account of the Bedford Park Suburb and Leopold Bloom's imagination of an alternative suburban lifestyle demands further consideration of the politics of the garden city movement in turn-of-the-century Ireland.

Initiated by Sir Ebenezer Howard's *To-morrow: a Peaceful Path to Real Reform* (1898)[71], the garden-city urban planning scheme was first popularised in England, and it soon started to impress the Irish reformers with its community design, especially during the period in which Home Rule sentiment was at its peak. However, it was also the time in which Irish Revivalists engaged in fervent activities to resist constitutional nationalism. Whereas these Revivalists promoted independent national industries to resist monopolizing British capital in the market, their anti-British, and hence anti-materialistic and anti-capital, attitude is ambivalently reflected in the Anglicised garden-city programme for which some of the Revivalists, as exemplarily voiced by AE in the *Irish Homestead*, propagandized. In an issue of the *Irish Homestead* released on January 15, 1910, AE devoted a substantial space in his column promoting the movement of garden cities. Under the title of "Garden Cities," AE writes,

> .... One of the most encouraging signs in modern industry is the emergence of great captains of industry who look upon the industrial enterprises they control less as a means of making huge fortunes than as an opportunity of creating industrial communities working under the happiest and healthiest circumstances. We have all heard of the garden cities at Port Sunlight, the Bournville Garden Village, the Earswick Garden Village, associated with great philanthropic manufacturers. America has its garden cities and its co-partnership enterprises where labour is being placed in a position to assume at first partial and gradually final control over the works which employ it. *We have no doubt these nobler ideals of labour united with capital will gradually extend from manufacturer to manufacturer, and will do much to ease off the friction between employer and the employed.* It has been found that the workmen in such enterprises labour more contentedly, with less friction, and this hearty concurrence of labour with capital does much to promote the success of the undertakings. Ireland has not yet had its garden cities or its co-partnership enterprises, but now that the co-operative idea is getting familiar to Irishmen and is inspiring so many workers, we expect it will begin to permeate industrial undertakings, and we may find some of our foremost industrial enterprise run on co-partnership lines. [...]. *The founders and workers in such communities are pioneers of the true brotherhood of humanity towards which everything is tending, and we would like to see Ireland doing its share of the noble pioneer work.*[72]

Indeed, the garden city's ideal of co-partnership communalism is founded in tune with the co-operative spirit which AE, Plunkett, and other members of the co-operative movement looked up to. AE's political aspiration, first as an organizer of the IAOS, then as editor of *The Irish Homestead* from 1906 on, is "to create a Celtic rural communalist civilization that ... [would] transform an economic into an ideological movement, promoting the establishment of local libraries and Irish literature as well as modern methods in the productions, marketing and distribution of dairy produce"[73]. Together with Plunkett's IAOS-founding principle to "regenerate Irish rural life by promoting a scientific agriculture organized on communalist self-help lines"[74], their agricultural co-operative credo echoes the communal idealism of the garden city advocates from Ebenezer Howard, Raymond Unwin, to Patrick Geddes and Patrick Abercrombie.

Howard's Garden City scheme outlined in *Tomorrow: A Peaceful Path to Real Reform* (later *Garden Cities of Tomorrow*) was very much a product of contemporary social regeneration movements in England, significantly influenced by William Morris and John Ruskin's call for community-conscious crafts production against industrialism. He was also immensely influenced by the Russian communalist anarchist Peter Kropotkin, whose series of writings that appeared in *The Nineteenth Century* magazine from 1888 to 1890 (these were later to become the book *Field, Factories, and Workshops,* a copy of which Joyce possessed) inspired his practical garden-city plan, designed after Kropotkin's "industrial villages" with electrical power and other modern technical appliances[75]. In admiration of Kropotkin, Howard even praises him as "the greatest democrat ever born to wealth and power."[76] Indeed, Howard shares with Joyce an indebtedness to socialist thought: whereas in "Circe" Bloom "stands for the reform of municipal morals and plain ten commandments" (*U* 15.1685), Howard's self-explanatory subtitle for his 1898 original edition of the book -- "*A Peaceful Path to Real Reform*" — says it all. In this book, he constantly refers to passages from socially conscious writers, including George Bernard Shaw, John Ruskin, and above all William Blake, whose work Joyce also constantly alludes to in *Ulysses*.

In the opening to his substantial first chapter of *Garden Cities of To-morrow*, Howard quotes from Blake:

> I will not cease from mental strife,
> Nor shall my word sleep in my hand,
> Till we have built Jerusalem
> In England's green and pleasant land.[77]

Readers of *Ulysses* will be reminded of Bloom's phantasmagorical speeches upon the foundation of New Bloomusalem in "Circe," where he "stands for the reform of municipal morals" (*U* 15.1685) and greets John Howard Parnell, Charles Stewart Parnell's brother, with gratitude for the latter's "right royal welcome to green Erin, the promised land of our common ancestors" (*U* 15. 1517–8). Similar discourse that evokes Ireland as the Promised Land and Parnell as the leading prophetic figure, a metaphor Joyce constantly uses since his 1912 journal article "The Shade of Parnell," recurs as one of the dominant comparisons between Ireland and the Promised Land, Parnell, and Moses, throughout *Ulysses*[78]. However, Howard was of course not the only socialist idealist of the time to associate his renovation ideal with Blakean Utopia. Raymond Unwin, the most influential British architect during this period, who was significantly the one to put Howard's garden city ideal into practice, elicits a similar vision in his conception of social regeneration through urban and suburban planning projects. The architectural showpieces he had undertaken exemplify each of the three strands that comprised the garden city movement: the industrial village, the garden city, and the garden suburb[79]. In Swenarton's study of this period's urban construction programmes and a shared socialist ideal in social and architectural movements, he indicates:

> [Raymond Unwin] looked to a millenarian transformation in society that was spiritual, ethical and communitarian all in one. In effect, although not in so many words, this

was the arrival on earth of the New Jerusalem foreseen by Saint John in the Book of Revelation, which had inspired a lineage of English millenarians and visionaries from the middle ages through to William Blake. [...] for Unwin commerce and convention were the enemies; what was needed was a rupture with convention that would both encourage and express a new way of living, in which a spirit of co-operation would replace competition and humanity would again live in harmony with nature.[80]

This co-operative vision, shared by AE in *The Irish Homestead* and Joyce's "Bloom of Flowerville" in "Ithaca," would be the foundation on which Ebenezer Howard and Raymond Unwin's New Jerusalem, AE's "Co-operative village," and indeed Leopold Bloom's "New Bloomusalem," were envisioned. In this respect, Joyce's Exodus metaphor indeed concurs with that of Howard and Raymond's socialist prospect and the Blakean vision, which the latter put forward concerning the construction of a Promised Land after the image of a New Jerusalem.

In *Garden Cities of To-morrow*, Ebenezer Howard puts forth designs for a "social city" that attempted to forge a bridge between the capitalist system and the socialist ideals that were gaining political impetus, with Trade Unions, Co-operatives and, central to Howard's argument, ideas of communal land protection. At the heart of his garden city scheme is the town-country magnet (see Fig 1.2) which, with a triangular diagram that outlines the appeals and downsides of either town-living or country-living lifestyles, explains his prospect of the Garden City as an in-between option that would incorporate advantages from both sides and avoid disadvantages from either. His Garden City would be "a marriage of town and country, of rustic health and sanity and activity and urban knowledge, urban technical facility, [and] urban practical co-operation"[81]. Following his predecessors, Howard sees in the metropolis of London an aggravated example of deteriorating living conditions as a result of the rise of capitalism and urbanization. He was, however, not alone in spotting the social problems of London in his time. In an issue of *The Daily Chronicle* released on November 6th, 1891, Sir John Gorst sensationally claims: "If they wanted a permanent remedy of the evil they must remove the cause; they must back the tide, and stop the migration of the people into the towns, and get the people *back to the land*"[82]. In the introduction to *Garden Cities of To-morrow*, Howard refers to Sir John Gorst's statement above and claims that, were his three-magnet blueprint put into practice, the uneven distribution of population and civil development between the country and the city and its modern social problems would be avoidable. He writes:

> The construction of such a [Town-country] magnet, could it be effected, followed, as it would be, by the construction of many more, would certainly afford a solution of the

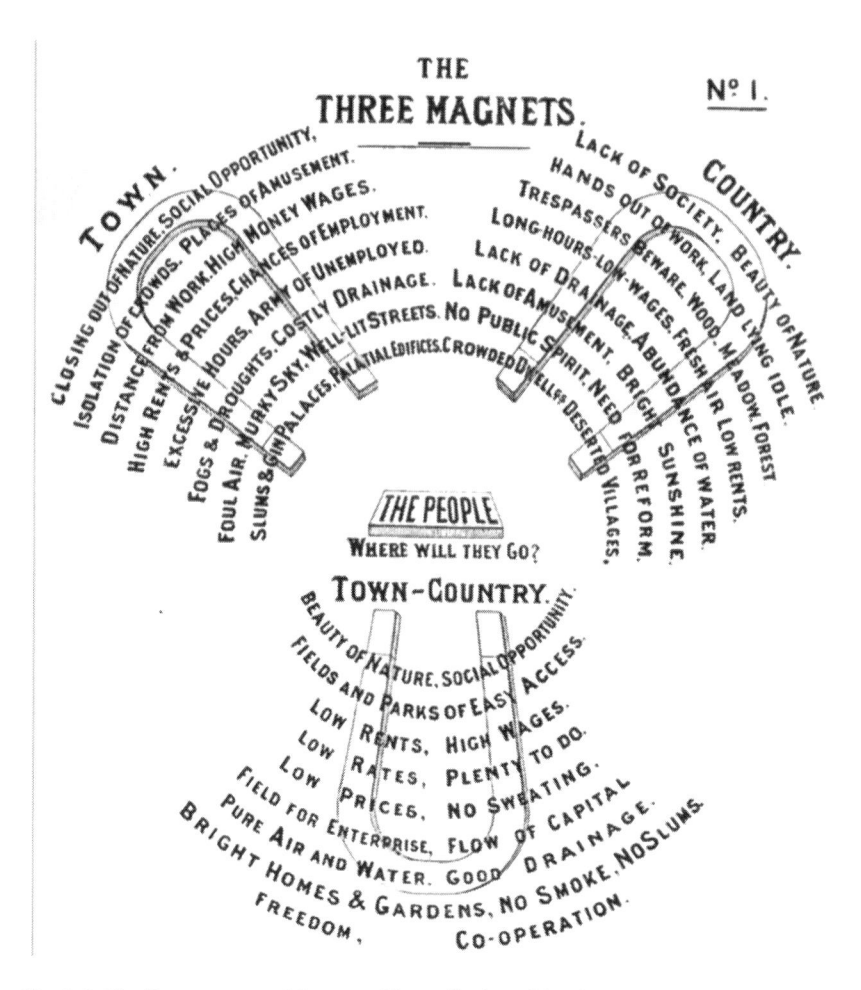

Fig. 1.2: The Town-country Magnets. Howard's idea of the three town-country magnets, illustrating what he perceives as the pros and cons of town and country livings, and ultimately, the advantages of a town-country, namely his garden city model. Source: Ebenezer Howard, *Garden Cities of To-morrow*.

> burning question set before us by Sir John Gorst, 'how to back the tide of migration of the people into the towns, and to get them *back upon the land*.'[83]

Curiously, in the Cabman's shelter while Leopold Bloom is reflecting on the land question and the eviction of tenants, he confesses that he "at the outset in principle at all events was in thorough sympathy with peasant possession as voicing the trend of modern opinion [...] and even was twitted with going a step farther than Michael Davitt in the striking views he at one time inculcated as a *backtothelander* [...]" (*U* 16.1593, my italics). Not only does Bloom recollect the call to return to the land to solve the contemporary social problems of Ireland, the "back-to-the-land" notion further reminds him of his earlier conflict with the Citizen in Barney Kiernan's pub, thus associating his political attitude toward the Land question with the novel's most notable, and perhaps most violent, scene of radical Nationalistic confrontation. Indeed here Bloom has moved farther than sympathising with Michael Davitt's agrarian socialism — his own claim is also a contemporary reverberation of the urban planners' call for social regeneration that was first concretised in Ebenezer Howard's *Garden Cities of To-morrow* and then put into practice in Raymond Unwin's urban planning projects from the industrial village in Rowntree's New Earswick York (1902), Letchworth Garden City (1903), to Hampstead Garden Suburb (1905) in suburban London.

For the socialist advocate of Garden Cities, the call to return to the land is conditioned by the securement of an agricultural estate for communal use. Howard spends much of the content of *Garden Cities of To-morrow* explaining and outlining the financial expenditures for this prospective estate, as is suggested by the title of its second chapter: "The Revenue of Garden City, and How it is Obtained — the Agricultural Estate." He believes that "the only difference between [his] proposal and the proposals as to land reform put forward in this book is not a difference of system, but a difference [...] as to the *method* of its inauguration"[84]. Instead of adopting Herbert Spence's advocacy of the more radical social transformation in the agricultural hierarchy, Howard proposed to "purchase the necessary land with which to establish the system on a small scale, and to trust to the inherent advantages of the system leading to its gradual adoption"[85]. The act of land acquisition by means of monetary transaction hence positioned the Garden City movement significantly as a social regeneration programme at the turn of the century.

The years 1900–30 saw a period in which the architecture, housing, and politics of modern Britain were formed. Housing was, according to Swenarton, "one of the main planks of social democratic politics (which in Britain meant Labour)"[86]; and in response to its emergent importance in social development, it inevitably became one of the areas in which anti-Labour political parties also desired to leave their mark. In the meantime, with its state-funded programmes, British architects saw in the advent of social democracy "both the opportunity and the necessity for a new kind of architecture," one which would act "as symbol and midwife of the new society emerging from the old"[87]. It was through these newly emerged architects' unprecedented claim "to expertise in the design for the working class" that they proclaimed a leading role in the social democratic pageant[88]. Under such pressure for social reformation and reconstruction from these socialist-minded architects, the Housing and Town Planning Act of 1919 was released as a considerable contribution to the development of social housing, along with the government's decision to provide Exchequer subsidies for working-class accommodation.

The Housing and Town Planning Act of 1919 is not only "a major development in social policy" as Wilding puts it[89], but is crucially and generally considered to be the major turning point in the history of public housing in Britain. Swenarton points out:

> Before the First World War the housebuilding activities of local authorities had been limited by the fact that any loss had to be borne by the rates; accordingly, the scope of those activities remained actively small, amounting to no more than five per cent of the housing built in any year. The Act of 1919 marked a break with previous policy by transforming the optional power of local authorities to provide housing into a duty, and by providing a Treasury grant to absorb any losses in the House Revenue Account in excess of those that could be met by a local rate of a penny in a pound.[90]

Bearing in mind the above-mentioned background of the 1919 Housing and Town Planning Act and the Act's fundamental role as a sign of unprecedented state intervention in financing Britain's social housing, it is curious to reconsider Joyce's change of wording from "The Irish Civil Service Building Society" (*U-G* 1582) to "Industrious Foreign Acclimatised Nationalised Friendly *Stateaided* Building Society" (*U* 17.1658–9, my italics) to be involved in Bloom's proposed financial scheme for his prospective residence of Bloom Cottage. In Joyce's later revision Bloom's possible funding source is not simply concerned with the hopeful aid of the state, but further refers to a "nationalized" institution that is ironically "foreign acclimatised" (*U* 17.1658–9). In this period of modern Ireland in which the concept of the "nation" has been under serious debate, Joyce's use of the terms "nationalised" and "foreign" appears controversial given the movement's political aims, especially during the turbulent time when it was introduced into Ireland.

Indeed, despite the British fervour for garden suburban planning since the late nineteenth century and AE's attentive promotion of Howard's Garden City campaign in his agricultural co-operative magazine, the Irish public barely paid the movement serious thought, mostly due to post-Parnellite political controversy and the Home Rule debate. The campaign's official introduction to Ireland was the launch of the Housing and Town Planning Association of Ireland (HTPAI) in September 1911 under the leadership of Lady Aberdeen, wife to

Lord Aberdeen, who was Lord Lieutenant of Ireland at the time. The Aberdeens were first exposed to the Garden City Movement when they visited Patrick Geddes' City and Town Planning Exhibition in Edinburgh earlier in the same year.[91] Geddes' belief — "by recapturing the lost values of citizenship, which he believed had existed in the pre-industrial city, there would be a regeneration of urban life which would sweep away the unhealthy industrial cities and the blind values of political nationalism" — immediately captured the Gladstonian Liberalist couple's attention[92]. They twice managed to bring Geddes' exhibition to Ireland, and later, as a result, organized HTPAI with the aim to "extend British town planning legislation to Ireland [...] and [to carry out] a civic survey in Dublin"[93]. With a belief in the prospect of combining "Geddesian ideas about civic duty and the construction [...] of Unwinian garden suburbs" to "replace the chronic slums in Irish cities"[94], the association was the major force behind the 1914 Dublin Town Plan Competition, an open competition to put the urban regeneration of Dublin into practice. The outcome of this competition attracted substantial attention from the press, if not so much from the Irish general public. The winning plan by Patrick Abercrombie received praise from the English architectural magazine *The Builder* and *Architectural Review*, the American Architectural magazine *Architect*, as well as the Irish newspaper *The Freemans Journal*[95], of which Leopold Bloom, who reads it in the "Calypso" episode of *Ulysses*, not only is a devoted reader but also serves as an advertising canvasser for the paper. In an issue of the *Times* on September 9th, 1916, Thomas Mawson even claimed: "it was essential that [Abercrombie's] plan be implemented in the wake of the Easter Rising, because ... 'the present occasion marks a crisis in the history of the town planning movement'"[96]. It is clear, then, that contemporary English, American, and Irish reviewers were impressed with the foresight of Abercrombie's regenerating civic planning and its relevance for Dublin and its environs, as well as how they recognized its timely contribution to the coeval town planning movement in Ireland.[97]

Composed and elaborated in the post-Easter-Rising period between spring 1921 and late January 1922[98], Bloom's Utopian suburban dream in "Ithaca" interestingly echoes Abercrombie's 1914 award-winning regeneration town plan for the city of Dublin. Some of the key entry requirements of the 1914 Dublin Town Plan Competition brief were to "provide the 14,000 new dwellings recommended by the Dublin Housing Inquiry ... [and] to be supplemented by a wholesale reorganization of the transport system and land use in Dublin and its environs"[99]. The winning architect Abercrombie drew his planning scheme largely from Howard's and Unwin's Garden City tradition, with the belief that "housing and transport links had to be reformed together on a grand scale if Dublin was to be rescued from economic stagnation"[100]. Likewise in "Ithaca," to follow up his more domestic Flowerville scheme, Leopold Bloom envisions a "wider-scoped" civic blueprint that shares identical concerns with Abercrombie's urban plans, with elaborative details on

> [a] scheme to be formulated and submitted for approval to the harbour commissioners for the exploitation of white coal (hydraulic power), obtained by hydroelectric plant at peak of tide at Dublin bar or at head of water at Poulaphouca or Powerscourt or catchment basins of main streams for the economic production of 500,000 W. H. P. of electricity. A scheme to enclose the peninsular delta of the North Bull at Dollymount and erect on the space of the foreland, used for golf links and rifle ranges, an asphalted esplanade with casinos, booths, shooting galleries, hotels, boardinghouses, readingrooms, establishments for mixed bathing. A scheme for the use of dogvans and goatvans for the delivery of early morning milk. A scheme for the development of Irish tourist traffic in and around Dublin by means of petrolpropelled riverboats, plying in the fluvial fairway between Island bridge and Ringsend, charabancs, narrow gauge local railways, and pleasure steamers for coastwise navigation (10/- per person per day, guide (trilingual) included). A scheme for the repristination of passenger and goods traffics over Irish waterways, when freed from weedbeds. A scheme to connect by tramline the Cattle Market (North Circular road and Prussia street) with the quays (Sheriff street, lower, and East Wall), parallel with the Link line railway laid (in conjunction with the Great Southern and Western railway line) between the cattle park, Liffey junction, and terminus of Midland Great Western Railway 43 to 45 North Wall, in proximity to the terminal stations or Dublin branches of Great Central Railway, Midland Railway of England, City of Dublin Steam Packet Company, Lancashire and Yorkshire Railway Company, Dublin and Glasgow Steam Packet Company, Glasgow, Dublin and Londonderry Steam Packet Company (Laird line), British and Irish Steam Packet Company, Dublin and Morecambe Steamers, London and North Western Railway Company, Dublin Port and Docks Board Landing Sheds and transit sheds of Palgrave, Murphy and Company, steamship owners, agents for steamers from Mediterranean, Spain, Portugal, France, Belgium and Holland and for Liverpool Underwriters' Association, the cost of acquired rolling stock for animal transport and of additional mileage operated by the Dublin United Tramways Company, limited, to be covered by graziers' fees. (*U* 17.1710–43)

Notice Bloom's schemes are mainly concerned with energy resources and transportation. Considering the language in this specific episode, the scientific discourse outlines a technical merchandise plan according to civic needs. Here by signifying the civic requirements according to his contemporary scientific and technical knowledge, Joyce is putting together a discourse that is not simply academic and scientific, but also practical and yet forward-looking. His attention to transportation points to modern civic traffic problems as a result of urbanisation, and echoes in practicality the 1914 Dublin Town Plan Competition's significant emphasis on a feasible solution to civic transportation and commodity circulation.

In spite of the correspondences between Abercrombie's Garden Suburban plan and the civic prospect outlined by Bloom of Flowerville, in the political climate in Ireland at the time, town planning could by no means offer a solution to the opposition of Nationalists and Unionists over the question of Home Rule. Indeed, despite the efforts of organizations like the Citizens Housing League, "there was no real integration of radical labour groups into the town planning movement in Dublin"[101]. As a result, Garden suburb ideals remained in the hands of "a rather narrow clique of middle-class reformers," and "[f]or adherents of the Irish Party and other Nationalist groups, any attempt to turn away from the question of independence was seen ultimately as only playing into the hands of Unionists and the British Government"[102]. Murray Fraser further indicates: "As has been observed of Sir Horace Plunkett, whom [Patrick] Geddes so much admired, the aim to bypass nationalism was perceived as simply another form of 'constructive Unionism' - that is, of promoting Irish social and economic reform to help divert from the demands for self-government"[103]. Not surprisingly, Garden City propaganda could not but be received by the radical supporters of Irish Nationalism as another diversion to the *real* solution to the Irish question. Due to the controversial political aspects of the Garden City movement in Ireland, no wonder Leopold Bloom, who earlier that day just stood against the Nationalist views of the Citizen in Barney Kiernan's pub (in the "Cyclops" episode), sticks to his fantasy of a prefigured Howardian suburban planning dream.

IV.

> *It is because we are thrown into history that we must cultivate our garden. .... Our human gardens may appear to us like little openings onto paradise in the midst of the fallen world, yet the fact that we must create, maintain, and care for them is the mark of their postlapsarian provenance. History without gardens would be a wasteland. A garden severed from history would be superfluous.*
>
> — Harrison, *Gardens: an Essay on the Human Condition*, p. x

It is important to recognise that Leopold Bloom's conception of and, indeed, his consumerist desire for "Bloom Cottage" of "Flowerville" isolates the house as an object — not simply from a consumerist perspective as a result of material culture, but also from the cultural perspective whence the modernist anxiety about contemporary environmental transformation derives. Such anxiety mediates between the contradictory country complex — in which the countryside was celebrated as an idyllic manifestation of a national repose yet its development after the Industrial Revolution has aroused debates over the question of urbanization and rural reconstruction. Raymond Williams indicates that the important eighteenth-century tradition of house-building and landscape-gardening was designed to fulfill "the new morality of improvement" to "reshape" and "redesign" the countryside[104], yet due to the impact of industrialization in the following century, the relationship between the city and the country witnessed enormous transformation. Rail transportation, mass production, and the rapid rise of capitalism all added to the controversial image of the idyllic countryside. The national image is often consciously invoked in this kind of bucolic narrative of the country, while the degeneration of the urban environment in industrial cities constitutes an alternative narrative of the nation. During the late nineteenth century, however, the emphasis of English poets and novelists shifted from "using the countryside as a device for criticizing the towns to an anxious concern about or pessimistic awareness of the threatened status of the former"[105]. A few decades later, the effects of the First World War further "reinforced this literary awareness of the fragility of rural England"; as a result, "more than ever the countryside became a place to retreat to (often in the imagination) from an unacceptable or unbearable contemporary world"[106]. On a practical level, to counter this civil problem, urban planners and architects initiate, mostly with socialist enthusiasm, the garden city movement, and its industrial model villages.

Interestingly, the garden city movement's socialist aspirations aside, it is ideologically indebted to a cultural tradition in which the hierarchical order is celebrated. The renowned architect of the Garden City campaign, Raymond Unwin, devotes himself to a vision of "organic unity that he believed had existed in the hierarchically arranged agricultural villages of the past"[107]. His belief that he could "pour the new wine of equalitarian socialism into the old bottles of hierarchical community"[108], however, does not in effect deprive his ideal socialist community of material relations under capitalism. In his comprehensive study of garden cities, architectural historian Walter Creese observes how both Ebenezer Howard and William Morris "tried […] exception to the anarchy of uncontrolled urban growth by a single grand gesture[:] Howard wished to exercise final power through the municipal ownership of land, while Morris hoped to restore the primacy of art in determining the environmental quality and texture of the community"[109]. What Creese suggests is that the two reformers' respective social regeneration plans both provide visions of a renewed civic relationship by proclaiming alternative ownership apart from mainstream capitalism. In the case of Howard a communal possession of land is envisioned; in the case of Morris, a revived material culture influenced by Arts and Crafts movement is projected. In both cases, as in that of Bloom Cottage, the desired ownership of property as well as of commodities are being righteously supported under the utopian ideal of a socialist communal structure. Speaking on urbanization and consumer culture, the social anarchist thinker Murray Bookchin comments how "modern city planning, by unconsciously assimilating commodity relations as social ideals, has [...] helped to produce designs that debase the city to a marketplace and raised structures that have turned it into the home of concentrated bureaucratic power"[110]. To further his point, he argues that as a consequence, in the modern city "the lack of consciousness becomes a form of consciousness, and the opportunism of technical success as a goal in itself degrades urban life precisely to the degree that technique celebrates its power to control the city's destiny"[111]. With Leopold Bloom's Flowerville model in "Ithaca," Joyce's ambiguous portrayal of Bloom's socialist inspirations, his material desire in the midst of commodity culture, and his civic ideal accompanied by technical advancement, are after all revealing in their own light of the complex urban situation of the modern city as Murray Bookchin outlined.

Regardless of its direct Edenic allusion to an earthly paradise, the garden often denotes an escapist utopia aspect from contemporary political turbulence, but "unlike earthly paradises, human-made gardens that are brought into and maintained in being by cultivation retain a signature of the human agency to which they owe their existence"[112]. Leo Marx's *The Machine in the Garden* (1964) sets the tone for a contemplation of garden politics that is to radically culminate in Howard's Garden City project. He deems a garden as "a miniature middle landscape. It is a place as attractive for what it excludes as for what it contains"[113]. Disguised in its prelapsarian archetypal form, the garden often affirms a human being's anxious discontent with his contemporary social condition. In "Gardens of Power and Caprice" (1984), Yi-Fu Tuan also speaks of the garden as "an artificial world made from the stuff of the real, the preeminent act of human will"[114]. Such a utopian miniature of a sublime Nature is what enables people to envision a community "in the image of a garden, an ideal fusion of nature with art," with which the landscape would become "the symbolic repository of value of all kinds — economic, political, aesthetic, religious"[115]. Whereas rural land and gardens are previously preserved for elite leisure, such undertones embedded in gardening would become even more obviously involved in politics during mid- to late nineteen-century England, when "gardens and allotments were a practical (if modest) realization of the deep-seated working-class demand for land whose political expression was radical agrarianism"[116]. Although the succeeding land acts made gardening more prevalent among working-class people, the practice of gardening still reflects a division in a social structure that has roots in class distinction and hierarchy, even to this day[117].

On the exclusiveness of power in gardens, Dean MacCannell comments: "What has been excluded from the garden is the *other*, not merely *an* other or the Other who watches over us but the radically other, that is, classes and ethnicities that are not allowed in the garden"[118]. As George McKay puts it in other terms, "[c]lass position is expertly articulated and affirmed through horticultural experience [...]. These open, mixed class horticultural spaces were about confirming and controlling social hierarchy"[119]. In the same light, Bloom's aspirational, ascending social status successive to his identity as a gardener (*U* 17.1608–9) reveals his political anxiety as a middle-class Jewish Irishman in the strict "hierarchical orders" (*U* 17.1608) of fin-de-siècle society. By positing himself in an already privileged position as a gardener, his desired ascent into a higher social class is made manifest, yet even more so is his unconscious anxiety over his contemporary social position. Considering the ethnopolitical and sexual oppression he has suffered from the Citizen and Molly's affair with Boylan, his political campaign that involves "upholding of the letter of the law" against "all orotund instigators of international persecution, all perpetuators of international animosities" (referring to the Citizen and the scene at Barney Kiernan's pub), "all menial molestors of domestic conviviality," and "all recalcitrant violators of domestic connubiality" (implicitly suggesting Molly's extramarital affairs with Boylan) (*U* 17.1627, 1630–33) can be easily justified and possibly avenged in his prospective political influence to be obtained through the act of gardening and subsequent promotion in the social hierarchy.

While garden politics is inseparable from the political narrative of human relations with man-made society and nature, it also plays a role in pedagogical functions. Since as early as the Academy in Athens, founded on the basis of Plato's concept "to create an environment that was hospitable to love, and environment in which the life philosophy could flourish as if under the care and supervision of a dedicated gardener"[120], the idea of gardening has been interwoven with education. Yet the analogy between gardening and pedagogy does not simply lie on the ideological level; instead gardening is practiced as one of the central curricula in education, especially exemplified in Epicurus' Garden School. Epicurus believes that "the purpose of philosophy was not to rule the city but to enhance mortal life's potential for happiness, precisely by liberating happiness from its traditional connections to citizenship," and "[l]ike the garden, personal happiness calls for self-cultivation and *culture de soi*"[121]. Such an idyllic picture of leisured country life is also celebrated in the curriculum of the emerging public schools in England after 1850. The prevailing values of the landed elite used in the education of school boys, including a rising number of upper-middle-class boys, inculcated an ideology that looks down upon industry, commerce, and business[122]. Bearing in mind that such biased social values were promoted in contemporary public-school education, including Epicurean *culture de soi*, Joycean readers would be reminded of Bloom of Flowerville with his upper-class intellectual and recreational pursuits (*U* 17.1581-1602), as well as the scientific, pedagogical style of Q&As in the "Ithaca" episode.

In the introduction to *The Story of Gardening*, Martin Hoyles says: "Gardening is often seen as a 'natural' occupation, just as capitalism and patriarchy are assumed to be 'natural' conditions which cannot be changed"[123]. Yet in fact, just as capitalism and patriarchy each has its distinct discourse of politics and relations, gardening is also "part of a particular political ideology"[124]. In this light, it is not sufficient to read Leopold Bloom's alternative identity as "Bloom of Flowerville" simply as a reflection of his desired lifestyle by engaging in an emerging consumer culture. It further signals Joyce's contemporary civic concerns with urban planning, the country-city complex, education and social hierarchy, as well as more specific political campaigns including co-operative movements and the Garden City movement. As Walter Creese comments: "The garden city movement can never be properly assessed if it is regarded [...] as an effort directed solely toward the future or offered only as an alternative to modern metropolitanism. It included these aims, but its deeper harmony and greater success would depend upon how well it *reconciled* the past with the present, town with country, and agriculture with industry"[125]. Likewise, neither Bloom's self-image as a gentleman farmer of Flowerville (*U* 17.1579-1615) nor his New Bloomusalem thoughts on "[e]mbellish[ing] (beautify[ing]) suburban gardens" (*U* 15.1667) is simply a fantasy or an alternative country lifestyle. It is rather an illustration of reconciliation of contemporary ideologies, politics, and modern civic problems. Indeed, when Thomas Jefferson confessed "No occupation is so delightful to me as the culture of the earth, and no culture comparable to that of the garden"[126], he was not simply thinking of the American dream as a rural republic. It was rather the machine in the garden — the question of industry, politics, and the nation revealed in a natural setting — that was at stake. In a similar light, as "Bloom of Flowerville" (*U* 17.1581) is envisioned "without excessive fatigue at sunset amid the scent of newmown hay, ameliorating the soil, multiplying wisdom, achieving longevity" (*U* 17.1585-7), such a prophetic self-image of Leopold Bloom as a noble farmer calls for the question of the machine in the garden in the Irish context. With the interwoven dynamics of consumer culture, agrarian socialism, social anarchism, nationalism, as well as the context of the garden city movement revealed in Bloom's idyllic alternative identity as Bloom of Flowerville, Joyce directs his readers to an ecopolitical reconsideration of gardening and gardens. After all, just as Harrison says, "[h]istory without gardens would be a wasteland"[127], gardens speak far beyond the postlapsarian experiences of human government over Nature or miniatures of Nature through planning and development. They bear witnesses to intricate relations between history, land, nature, and politics, and, in the case of Flowerville, probably speak a more all-around word than those beautifully and ideally propagandized in AE's the *Irish Homestead* or Ebenezer Howard's *Garden Cities of To-morrow*.

Notes

1. An early draft of a section of this chapter was published in *Polymorphic Joyce*, the 12th issue of Joyce Studies in Italy, as "'Bloom of Flowerville': an Agrinational Consumer" (ed. by Franca Ruggieri and Anne Fogarty, Roma 2012, 73–83). I have since elaborated extensively on the materials in the original *JSI* piece, adding issues such as Garden City Movement and politics of gardening to my reading. I am thankful to the editors of *JSI* for allowing me to reprint the previously published materials here.
2. Vincent J. Cheng, *Joyce, Race, and Empire* (Cambridge: Cambridge UP, 1995), p. 241.
3. James Fairhall, *James Joyce and the Question of History*, (Cambridge: Cambridge UP, 1993), p. 4.
4. Declan Kiberd, *Ulysses and Us: The Art of Everyday Living* (Faber and Faber, 2009), p. 248.
5. Ibid.
6. Ibid., p. 255.
7. Among others, see Brian G. Caraher's discussion of the self/other dialectics in "A Question of Genre: Generic Experimentation, Self-composition, and the Problem of Egoism in *Ulysses*" (*ELH* 54.1, Spring 1987, p. 183–214), John Nash's *James Joyce and the Act of Reception*, and Clare Hutton, "Joyce and the Institution of Revivalism" (*Irish University Review* 33.1: 117-32) and her article in the collection *Joyce, Ireland, Britain* (2006), edited by Andrew Gibson and Len Platt.
8. John Nash, *James Joyce and the Act of Reception: Reading, Ireland, Modernism* (Cambridge: Cambridge UP, 2006), p. 84.
9. James Fairhall, "Northsiders," in *Joyce: Feminism/Post/Colonialism*, ed. by Ellen Carol Jones (Amsterdam: Rodopi, 1998), pp. 43-80.
10. Ibid., p. 48.

11. Jennifer Wicke, "Joyce and Consumer Culture," in *The Cambridge Companion to James Joyce*, ed. by Derek Attridge (Cambridge UP, 2004), p. 251.
12. Ellen Carol Jones, "Commodious Recirculation: Commodity and Dream in Joyce's *Ulysses*," *Joyce and Advertising*, eds. by Garry Leonard and Jennifer Wicke, Special Issue, *James Joyce Quarterly* 30.4-31.1 (Summer/Fall 1993), p. 745.
13. Cheng, p. 219.
14. Don Gifford and Robert J. Seidman, *Ulysses Annotated: Notes for James Joyce's Ulysses* (Berkeley: University of California Press, 1974), p. 479.
15. Cheng, p. 221.
16. Joseph Lee, *Ireland: 1912-1985* (Cambridge: Cambridge UP, 1989), p. 72.
17. For more details on the economic history of Irish land problem, see Michael Turner, *After the Famine: Irish agriculture, 1850-1914*. F.S.L. Lyons' *Ireland Since the Famine*, on the other hand, offers a more comprehensive study on the economical political background of the Irish land problem.
18. Michael Turner, *After the Famine: Irish agriculture, 1850-1914*, (Cambridge: Cambridge University Press, 1996), p. 216.
19. For an in-depth study on Phoenix Park Murder, see James Fairhall's *James Joyce and the Question of History* (Cambridge: Cambridge UP, 1993), chapter one in particular.
20. Gifford, p. 592, my italics.
21. Ibid., p. 558.
22. Ibid., p. 558.
23. Fairhall, *James Joyce and the Question of History* (Cambridge: Cambridge UP, 1993), p. 74.
24. Fairhall, *James Joyce and the Question of History* (CUP: 1993), p. 74.
25. P. J. Mathews, *Revival: The Abbey Theatre, Sinn Féin, the Gaelic League and the Co-operative Movement* (Cork: Cork UP, 2003), p. 29.
26. P. J. Mathews, "'A.E.I.O.U.': Joyce and the *Irish Homestead*," *Joyce on the Threshold*, eds. by Anne Fogarty and Timothy Martin (Gainesville: University Press of Florida, 2005), p. 161.
27. F.S.L. Lyons, *Ireland Since the Famine*, (Fontana Press, 1985), p. 210.
28. Desmond Roche, *Local Government in Ireland* (Dublin: Institute of Public Administration, 1982), p. 46.
29. P. J. Mathews, *Revival: The Abbey Theatre, Sinn Féin, the Gaelic League and the Co-operative Movement* (Cork: Cork UP, 2003), p. 8
30. Ibid., p. 8.
31. "Since the hewing down of that great overshadowing tree [Parnell] other growth have had a chance of stretching towards the sunlight, and new forces, the Society of Agricultural Co-operation and the Gaelic League, that I will try [sic], in this quiet moment, to show the character of Sancho-Quixote" (Gregory, "Ireland, Real and Ideal" 770, qtd. in Mathews 151).

32. John Hutchinson, *The Dynamics of Cultural Nationalism: The Gaelic Revival and the Creation of the Irish Nation State* (Allen & Unwin, 1987), p. 178, italics mine.
33. Mathews, *Revival*, p. 3.
34. This dichotomy, as well as a socio-historical outline of the cultural nationalist politics in Ireland, was initially put into lengthy study in John Hutchinson's influential book *The Dynamics of Cultural Nationalism* (London: Allen & Unwin, 1987).
35. Mathews, *Revival*, p. 7.
36. See Len Platt's *Joyce and the Anglo-Irish: a Study of Joyce and the Literary Revival*, and Andrew Gibson's *Joyce's Revenge: History, Politics, and Aesthetics in* Ulysses. Clare Hutton's articles on Joyce and the Revival also constructively inform the cultural politics and historical background of the Anglo-Irish Revival.
37. Len Platt, "'If Brian Boru Could But Come Back And See Old Dublin Now': Materialism, the National Culture And *Ulysses* 17," in *Joyce's Ithaca*, ed. by Andrew Gibson (Amsterdam: Rodopi, 1996), p. 113.
38. Ibid., p. 122.
39. Leeann Lane, "'It is In the Cottages and Farmers' Houses that the Nation is Born': AE's *Irish Homestead* and the Cultural Revival," in *Irish University Review* 33.1 (Spring/Summer 2003), p. 166.
40. Ibid., p. 165.
41. George W. Russell (A.E.), *Selections from the Contributions to the Irish Homestead*, Vol. 1-2, ed. by Henry Summerfield (London: Colin Smythe, 1978), p. 50.
42. Ibid., p. 50.
43. Hutchinson, p. 81.
44. Petrie was co-editor of the *Dublin Penny Journal* (1832–3), and editor of the *Irish Penny Journal* (1840–1), the publication of which especially popularised the construction of a "regenerated Irish national culture" (Hutchinson 80–90).
45. Hutchinson, p. 89, my italics.
46. A comment made by J. M. Synge. See A. Price, ed., *J. M. Synge: Collected Works, Vol. II: Prose* (London: Oxford University Press, 1967), 283.
47. Lane, p. 167, my italics.
48. Platt, *Joyce and the Anglo-Irish*, p. 183.
49. Hutchinson, p. 132-4.
50. Hutchinson, p. 184.
51. Arthur Griffith, *The Resurrection of Hungary: A Parallel for Ireland.* (1904), with introduction by Patrick Murray (Dublin: University College Dublin Press, 2003), p. 139–63.
52. Dominic Manganiello, *Joyce's Politics* (London: Routledge & Kegan, 1980), p. 124.

53. George W. Russell (A.E.), *Co-operation and Nationality: A Guide for Rural Reformers from This to the Next Generation* (Dublin: Maunsel & Company, 1912), p. 50.
54. Nicholas Allen, *George Russell (AE) and the New Ireland* (Dublin: Four Courts, 2003), p. 46.
55. For Joyce's involvement in Irredentism and relevant socialist politics, see chapter two, "Young Ireland," of Dominic Manganiello's *Joyce's Politics*, as well as Brian G. Caraher's "Cultural Politics and the Reading of 'Joyce': Cultural Semiotics, Socialism, Irish Autonomy, and 'Scritti Italiani'" (*JJQ* 36.2: 171–214).
56. Peter Kropotkin, *The Conquest of Bread* (London: Chapman & Hall, 1913), p. 14.
57. Ibid., p. 36.
58. Manganiello, p. 113.
59. Peter Kropotkin, *Fields, Factories, and Workshops* (London: Swan Sonnenschein, 1907), p. 201.
60. Kropotkin, *The Conquest of Bread*, pp. 201, 226.
61. Ibid., 226.
62. Jones, p. 740.
63. Žižek, p. 28, qtd. by Jones, p. 741.
64. Murray Fraser, *John Bull's Other Homes: State Housing and British Policy in Ireland, 1883-1922* (Liverpool: Liverpool UP, 1996), p. 320.
65. Ibid.
66. The significance of the Arts and Crafts movement to the contemporary cultural and political movement in Ireland can be referred from Jeanne Sheehy's comment that: "Just as the earlier movement towards national feeling in Irish art was linked to the Gothic Revival in England, so the Celtic Revival was associated with the Arts and Crafts movement" (152). The movement's main propelling organisation, the Arts and Crafts Society of Ireland, was founded in 1894 "with the objects of fostering artistic industries in Ireland, promoting artistic culture by means of lectures and the supply of designs, and holding exhibitions of Irish arts and crafts" (Sheehy 152).
67. Jeanne Sheehy, *The Rediscovery of Ireland's Past: the Celtic Revival 1830-1930* (London: Thames and Hudson, 1980), p. 134.
68. Vernacular architecture refers to the construction of buildings using local methods and materials according to regional climate, traditions and cultures. Such a spirit of regionalism corresponds well with the ideal of the Arts and Crafts Movement in late-nineteenth-century Ireland. For detailed information on vernacular architecture and its relation with modernism, see *Vernacular Modernism: Heimat, Globalization, and the Built Environment*, edited by Maiken Umbach and Bernd Hüppauf (Stanford: Stanford University Press, 2005). For reading on the particular neo-vernacular trend in the history

of Irish architecture, see chapter eight of Sheehy's *The Rediscovery of Ireland's Past*, pp. 121–45.
69. Sheehy, p. 141.
70. Qtd. in Joseph Hone, *W. B. Yeats, 1865-1939* (London: Mac millan & Co, 1943), pp. 26–27.
71. The book is later retitled and published as *Garden Cities of To-morrow* (1902).
72. Russell, *Selections from the Contributions to the Irish Homestead*, p. 863, my italics.
73. Hutchinson, p. 143.
74. Ibid.
75. Pater Hall, Dennis Hardy and Colin Ward, "Commentators' Introduction," in *To-morrow: a Peaceful Path to Real Reform* by Ebenezer Howard (London: Routledge, 2003), p. 8.
76. Quoted in Fishman's *Urban Utopias in the Twentieth Century: Ebenezer Howard, Frank Lloyd Wright and Le Corbusier* (New York, Basic Books, 1977), p. 37, from manuscripts of Howard's unpublished autobiography.
77. Qtd. by Howard, p. 12.
78. See Abby Bender's book *Israelites in Erin* (Syracuse UP, 2015) for an insightful in-depth study on this topic.
79. Mark Swenarton, *Building the New Jerusalem: Architecture, Housing and Politics 1900-1930* (Bracknell: IHS BRE Press, 2008), p. 4.
80. Ibid., p. 5.
81. Lewis Mumford, "The Garden City Idea and Modern Planning," *Garden Cities of To-morrow*, by Ebenezer Howard, ed. by F. J. Osborn (London: Faber and Faber, 1945), p. 34.
82. *Daily Chronicle,* 6th November, 1891, qtd. by Howard, p. 3, my italics.
83. Howard, p. 11, my italics.
84. Ibid., p. 123.
85. Ibid., p. 123.
86. Swenarton, *Building the New Jerusalem: Architecture, Housing and Politics 1900-1930* (HIS BRE Press, 2008), p. 1.
87. Ibid.
88. Ibid.
89. Paul Wilding, "The Housing and Town Planning Act 1919 -- a Study in the Making of Social Policy," in *Journal of Social Policy* 2.4 (1973), p. 317.
90. Swenarton, p. 13.
91. A *Guide-Book and Outline Catalogue* of the Cities and Town Planning Exhibition was published in the same year (1911), authored by Patrick Geddes and F. C. Mears (Dublin: Browne and Nolan, 1911). Interestingly, according to the content of the exhibition indicated by this guide-book, among the exhibition rooms, Room X specifically targets at a "Survey of Dublin."
92. Fraser, p. 133.

93. Ibid., p. 134.
94. Ibid., p. 134.
95. "Abercrombie's town plan for Dublin ... was described by H. V. Lanchester as 'the highest development that had yet been reached in the matter of town planning in the United Kingdom', and Raymond Unwin recommended that it be studied by 'town planners all over the world, because it showed the state to which town planning was brought' (Fraser 141).
96. Fraser, p. 142.
97. Abercrombie's award-winning town plan was eventually published in 1922 under the title *Dublin of the Future: the New Town Plan* (Liverpool: University Press of Liverpool, 1922), as the first of the several volumes of the publications of the Civics Institute of Ireland.
98. For more extensive information on Joyce's composition history for "Ithaca," see the detailed elaboration in the afterword of the critical edition of *Ulysses*, edited by Hans Walter Gabler, in volume three, pp. 1885–91.
99. Fraser, p. 138.
100. Ibid.
101. Ibid., p. 145.
102. Ibid.
103. Ibid.
104. Raymond Williams, *The Country and the City* (New York: Oxford University Press, 1973), p. 59.
105. Jeremy Burchardt, *Paradise Lost: Rural Idyll and Social Change since 1800* (London: I.B. Taurus, 2002), p. 9.
106. Ibid.
107. Standish Meacham, *Regaining Paradise: Englishness and the Early Garden City Movement* (New Haven: Yale UP, 1998), p. 3.
108. Ibid., p. 4.
109. Walter L. Creese, *The Search for Environment: the Garden City Before and After* (New Haven: Yale UP, 1966), p. 150.
110. Murray Bookchin, *The Limits of the City*, 2nd rev. ed. (Montreal: Black Rose Books, 1986), p. 147.
111. Ibid.
112. Robert Pogue Harrison, *Gardens: an Essay on the Human Condition* (Chicago: University of Chicago Press, 2008), p. 7.
113. Leo Marx, *The Machine in the Garden: Technology and the Pastoral Ideal in America* (London: Oxford UP, 1964), p. 138.
114. Qtd. in Robert B. Riley, "Flower, Power, and Sex," *The Meaning of Gardens: Idea, Place, and Action*, eds. by Mark Francis and Randolph T. Hester, Jr. (Cambridge: MIT Press, 1990), p. 61.
115. Marx, *The Machine in the Garden*, p. 228.
116. Burchardt, pp. 51–52.

117. Exemplified in particular in contemporary community gardens and guerrilla gardens in terms of labour and class division, as well as manipulative gardening politics via flower plots in Third World countries; see George McKay's extensive study, *Radical Gardening: Politics, Idealism & Rebellion in the Garden* (2011).
118. Dean MacCannell, "Landscaping the Unconscious," in *The Meaning of Gardens: Idea, Place, and Action*, eds. by Mark Francis and Randolph T. Hester, Jr. (Cambridge: MIT Press, 1990), p. 96.
119. George McKay, *Radical Gardening: Politics, Idealism & Rebellion in the Garden* (London: Frances Lincoln, 2011), p. 16.
120. Harrison, p. 69.
121. Ibid, p. 80.
122. Burchardt, p. 94.
123. Martin Hoyle, *The Story of* Gardening (London: Journeyman Press, 1991), p. 5.
124. Ibid.
125. Creese, p. 5.
126. Qtd. in Marx, *The Machine in the Garden*, p. 138.
127. Harrison, p. x.

2 WASTE
Joyce's "cloacal obsession" and the Eco-politics of Waste

> *The turd swiftsure*
> *Flew down the sewer*
> *& the sleuce-hounds*
> *Flushfleshed after*
> — James Joyce, *The Finnegans Wake Notebooks at Buffalo VI, B.3, 67–69*

> *Was vast wealth acquirable through industrial channels? The reclamation of dunams of waste arenary soil, proposed in the prospectus of Agendath Netaim, Bleibtreustrasse, Berlin, W. 15, by the cultivation of orange plantation and melonfields and reafforestation. The utilisation of waste paper, fells of sewer rodents, human excrement possessing chemical properties, in view of the vast production of the first, vast number of the second and immense quantity of the third, every normal human being of average vitality and appetite producing annually, canceling byproducts of water, a sum total of 80 lbs. (mixed animal and vegetable diet), to be multiplied by 4,386,035, the total population of Ireland according to census returns of 1901. (U 17.1698-1708)*

In "Ithaca," for his communal development plan of Flowerville, Leopold Bloom conceives a revolutionary industrial scheme to provide for resources necessary for the foreseeable expenses involved in his blueprint of civil and domestic pursuits. Among them is the utilization of waste, human excrement, and sewage resources, as well as a land reclamation plan in Palestine. While the first part of Bloom's industrial development design is based on his impression of the Zionist project[1], his environmental plan for Irish regeneration takes over as his mind moves forward to more detailed recycling tasks. Indeed, the idea of waste reclamation, together with a cultivation scheme by the edge of the Sea of Kinneret, was rising topics of debate at the time. P. Anderson Graham opens his socio-economic study *Reclaiming the Waste: Britain's Most Urgent Problem* (1916) bluntly with the following statement: "The reclamation of waste land is one of the most urgent needs of the hour"[2]. Such rapidly growing awareness of a need for a more sustainable environment for economic and agriculture development in Britain and Ireland has a significant impact on Joyce's portrayal of Bloom's private utopian vision and his depiction of the distribution of Palestinian land reclamation propaganda.

Starting from the concept of waste in the more general context of *Ulysses*, the first part of this chapter attempts to place "Agendath Netaim" against the background of the Zionist movement at the start of the twentieth century, particularly alongside an often-neglected book from 1916, *Zionism and the Jewish Future,* to contemplate the historical context of this collection that in fact had a significant influence on Joyce's utopian depiction of a wasteland-reclaiming scheme in Palestine. In the first section of this chapter I intend to trace the novel's environmental references to the "Agendath Netaim" wasteland reclamation scheme, and discuss Joyce's use of the leaflet for "Agendath Netaim" throughout *Ulysses* as a means to reveal the contradictions of a contemporary consumerist approach to an environmental movement that is religious, economic and political in nature. On the other hand, "waste" not only as materialistic compost but also as a cultural product of civilisation raises questions with regard to how the "Agendath Netaim" leaflet, the "Elijah is coming!" flyer, sewage, together with other forms of waste, compose an alternative modern narrative on/of Bloom's "DEAR DIRTY DUBLIN" (*U* 7.921). In the later parts of this current chapter, I will be focusing on the dynamics of the cultural politics of waste, with thoughts on infrastructure, civilization, and power. Waterways and sewers in Joyce's time were much-debated aspects of modernization, evidenced in the twenty-seven-volumed "Royal Commission on Sewage Disposal" (1902), which contains reports on the condition of waterways and sewage systems in Great Britain's major cities including London, Birmingham, Belfast, and, of course, Dublin. Indeed, as Cheryl Herr and Michael Rubinstein persuasively point out in their works[3], Joyce's attentiveness to sewage in *Ulysses* discloses an alternative "plumbing consciousness," or "sewer of consciousness"[4] of the narrative. Dominique Laporte's study on the politics of shit also throws provocative light on the controversial role sewers have to play in the empirical construction of power. As we reconsider the history and politics of sewage and sewers in light of fin-de-siècle Dublin, and more expressly as Joycean readers, we cannot but also take into reconsideration H. G. Well's comment on Joyce's "cloacal obsession": how Joyce would "bring back into the general picture of life aspects which modern drainage and modern decorum have taken out of ordinary intercourse and conversation"[5]. By doing so, Joyce invites his readers via the scatological experiences of everyday Dublin, through the "sewer of consciousness," to the eco-politics of waste, to which we will return later in this chapter.

I.

> *Where there is dirt there is system.*
> — Mary Douglas, *Purity and Danger*, 35

As a man of the world, Leopold Bloom is attentive to products — products either as the aims of desires or as the results of acts of consumption. From window shopping on Henry Street to Plumtree's potted meat and Epps's cocoa (among many other products) in Bloom's kitchen, *Ulysses* is both branded by objects and objectified by brands. As much as Joyce's Dublin is marked by mass culture and consumerism, it is equally marked by what seems to be the opposite of consumption — the need for disposal and disposables. On waste-polluted Sandymount beach, Stephen Dedalus spent part of his morning pondering over the historical dialectics of subsistence (*U* 3.150–57), while the excrement issue lingers in Leopold Bloom's thoughts from the "Calypso" episode to "Ithaca" (*U* 4. 475–84; 17.1699–1708).

In their challenging studies on garbage and culture, Gay Hawkins and Gillian Whiteley both indicate that the organization of waste is the performance of civilization that reveals the modern system of classification and relations. The history of waste itself, Whiteley states, "has been the history of separating organic human waste from the rest"[6], and the study of trash leads to "a fundamental link to systems of value which are time and place specific"[7]. In other words, garbage as the disposed product of culture points to a cultural system that is intrinsic to consumer culture. Indeed it would be impossible not to read waste in light of the cultural system and the modern hierarchy in which consumerist desire plays a significant role. It is the collection of the not-haves, the leftovers, the used, and the "valueless". Hawkins indicates that waste is "a social text that discloses the logic or illogic of a culture [and is] subordinated to human action, a slave to desire and manipulation"[8]. It reveals human subjectivity, albeit with an opposite perspective from acts of consumption, by the practice of material selections in cultural and historical context. In Hawkins' words, it is "what we want to get rid of [that] tells us who we are"[9]. Shoar and Stam beautifully put it:

> The trash of the haves become the treasure of the have-nots: the dark and unsanitary is transmogrified into the sublime and the beautiful ... as a diasporized, heterotropic site, the point of promiscuous mingling of rich and poor, center and periphery, the industrial and the artisanal, the organic and the inorganic, the national and the international, the local and the global: *as a mixed syncretic, radically decentred social text, garbage provides an ideal postmodern and postcolonial metaphor.*[10]

Whereas the language and organization of garbage speak of the modern system of classification and order as well as suggest the postmodern and postcolonial hierarchy inherent in the system of consumptions and disposals, in a way, they also demystify the connection between waste and culture in modernist writings and after. Whiteley notes that "cleanliness and hygiene [are] about re-ordering the environment with all the psychoanalytical implications of that activity"[11]. Quoting from a study of the history of Unilever by Anne McClintock, she writes how "[Unilever's] advertising campaign for *Imperial Leather* soap — with its slogan, 'Soap is Civilisation' — conveyed the idea that imperialism literally washed away the dirt of 'primitivism'" — in this sense cleanliness equated "not only with order but with the 'civilisation' of colonialism and empire"[12].

If the soap Bloom purchases for Molly and keeps in his pocket for most of the day in a similar sense speaks of the modern dichotomy between civilisation and primitivism, his every-so-often concern for the disposables — the throw-away, sewage, excrement, etc. — intensifies such consciousness of the opposition between the civilised and the primitive. Gay Hawkins brilliantly illustrates how human habits related to waste disposal constitute a human narrative of subjectivity:

> Waste doesn't just threaten the self in the horror of abjection, it also *constitutes* the self in the habits and embodied practices through which we decide what is connected to us and what isn't. [...]. This is why styles of waste disposal are also types of self and why waste management, in all its cultural mutations, is fundamental to the practice of subjectivity.[13]

Indeed, in a corresponding way Leopold Bloom's thoughts on waste disposal both in "Calypso" and "Ithaca" reveal the subjectivity of human relation to objects, consumption, and functionality. As we return to the first scene in which Bloom considers the productivity of sewage on his way via his garden to the water closet, we find his thoughts astoundingly organic:

> He bent down to regard a lean file of spearmint growing by the wall. Make a summerhouse here. Scarlet runners. Virginia creepers. Want to manure the whole place over, scabby soil. A coat of liver of sulphur. All soil like that without dung. Household slops. Loam, what this is this that is? The hens in the next garden: their droppings are very good top dressing. Best of all though are the cattle, especially when they are fed on those oilcakes. Mulch of dung. Best thing to clean ladies' kid gloves. Dirty cleans. Ashes too. Reclaim the whole place. Grow peas in that corner there. Lettuce. Always have fresh greens then. .... (*U* 4.475–84)

Bloom's domesticated thoughts on gardening here incorporate a rather organic interaction between gardening and sewage systems. Significantly, his interior monologue doesn't stop there. A few lines later, just before he "kick[s] open the crazy door of the jakes" (*U* 4.494) and enters, his thought is diverted to the Agendath Netaim leaflet which he retains earlier at the butcher's place: "Deep voice that fellow Dlugacz has. Agendath what is it? Now, my miss. Enthusiast" (*U* 4.492–3). The sewage "reclaim[ing] of the whole place" (*U* 4.482) and the Zionist reclaiming of the Jewish homeland hereby merge to share the same act of proclaiming authority over the land. The act of reclamation of the land, as a performance of demonstration of power and authority, is here contrasted in two diverse ways to ridicule the Zionist plantation scheme as well as to proclaim the "colonialization" of waste in the land as part of the ecosystem of modernization and consumption. The postcolonial and postmodern context in reading waste and disposables as objects of have-nots can hence be read in an opposite way: waste is not so much the leftovers of imperial consumerism but a substance of dominance that saturates the land and by such means makes its substantial land reclamation. In this way, just as the "Elijah is coming" flyer advertises its presence along River Liffey, one of the sewage outlets of the city[14], through Bloom's organic thoughts on waste and gardening, readers are able to spot an eco-political dynamic of the ecosystem — in which the whole place can be "reclaimed" by cattle dung, household slops, hen droppings, etc. (*U* 4.478–82). Later in the "Ithaca" chapter Bloom further considers the sewage and waste system in light of the profitable prospect of an economic/ecological civil plan:

> Was vast wealth acquirable through industrial channels?
> The reclamation of dunams of waste arenary soil, proposed in the prospectus of Agendath Netaim, Bleibtreustrasse, Berlin, W. 15, by the cultivation of orange plantations and melonfields and reafforestation. The utilisation of waste paper, fells of sewer rodents, human excrement possessing chemical properties, in view of the vast production of the first, vast number of the second and immense quantity of the third, every normal human being of average vitality and appetite producing annually, canceling byproducts of water, a sum total of 80 lbs. (mixed animal and vegetable diet), to be multiplied by 4,386,035, the total population of Ireland according to census returns of 1901. (*U* 17.1698–1708)

Here again, Joyce provides a parallel reading of the reclamation of Palestinian land through the Agendath Netaim scheme with the reclamation of waste in Ireland. It would be interesting to add here that land reclamation was at the time an emerging aspect of social propaganda in Britain and Ireland. A case in point is P. Anderson Graham's *Reclaiming the Waste* (1916), which opens aptly and sensationally with a chapter named "The Urgency of Land Reclamation." Throughout this socio-economic study, Graham draws out the agricultural benefits as well as the foreseeable economic and environmental advantages to be gained from wasteland reclamations. Although these agricultural schemes Graham proposes are not identical with the waste reclamation Leopold Bloom devises for Ireland as part of his Utopian domestic environmental plan, interestingly among Graham's wasteland reclamation schemes is his contention for afforestation in Ireland, which reminds us of John Wyse Nolan's proposal of afforestation in the episode of "Cyclops"[15].

Don Gifford points out that where "Agendath Netaim" has occurred throughout *Ulysses* it refers to the propaganda of "a company of planters" advertised for a Zionist colony in Palestine[16]. The aim of this company, as he quotes from Hyman, was "to save the prospective settler the initial hardships involved in setting up a farm by itself, buying land, developing it, and planting trees for him"[17]. Gifford also indicates a likely reference to the company's address shown on the Agendath Netaim leaflet, Bleibtreustrasse 34, Berlin, W. 15: "Bleibtreustrasse 34-35"[18], the actual address for The Palestine Land Development Company in Berlin. Not surprisingly, like the Agendath Netaim campaign, the Palestine Land Development Company is also closely associated with the Zionist movement. It was one of the Zionist institutions listed in the appendix to *Zionism and the Jewish Future* (1916), and it was specifically founded to "acquire and develop land in Palestine, and to resell it in small holdings to prospective settlers"[19].

Over the decades, significant Joycean research has been undertaken to demystify "Agendath Netaim" and its Zionist associations in *Ulysses*. Indeed, many brilliant studies[20] have focused on "the Jewish problem" and Joyce's attitude toward the rise of anti-Semitism all over Europe at the turn of the century. The emphasis of the following pages is not so much to dwell on the complexity of Bloom's Jewishness, which has been comprehensively explored by outstanding scholars over the years[21]. Rather, I would draw particular attention to *Zionism and the Jewish Future*, a book in Joyce's Trieste collection which strikingly had been mostly neglected in recent Joycean scholarship until the appearance of Abby Bender's study *Israelites in Erin* in 2015. Joyce himself owned *Zionism and the Jewish Future* (1916)[22], a contemporary collection of writings by a few Zionists with diverse perspectives on the Zionist movement, as part of his book collection as of 1920. It would be likely, to say the least, that he owes much to this book for his use of the Agendath Netaim material in *Ulysses*. Indeed, I would like to suggest that the diverse articles in this collection, in particular S. Tolkowsky's "The Jews and the Economic Development of Palestine," provide Joyce with materials to outline the Palestine Plantation scheme, as well as to characterise Bloom's identity as an Irish Jew, and furthermore to foreground his political attitude and some of his socio-economic solutions for his contemporary environmental issues.

One of the contributors to this collection, Tolkowsky, an agricultural engineer in Jaffa, mentions in his article entitled "The Jews and the Economic Development of Palestine" an agricultural development scheme established in 1908 for the farm of *Kinnereth* on the shores of Lake Tiberias[23]. Along with detailed descriptions of all the schemes involved in the different land reclamation plantations by the Palestine Land Development Company, Tolkowsky encloses a map of the developing Jewish colonies in Palestine (see Fig 2.1). This may be the model on which Joyce bases his Agendath Netaim, whose first appearance in the novel is introduced as "the model farm at Kinnereth on the lakeshore of Tiberias" (*U* 4.154–5). Indeed the map shows quite a few Jewish colonies already established in Palestine under the Zionist movement; among them, on the lakeshore of Tiberias, is one located at a place called Kinnereth. In Appendix III to this particular volume, which includes a comprehensive list of the Jewish colonies in Palestine as of its publication, two other colony names are listed under Kinnereth, making the total number of acres of reclaimed land in the Kinnereth area nearly 2800 acres.

Our reading of *Zionism and the Jewish Future* as a fundamental source for Joyce's composition of *Ulysses*, however, is not limited to the documentation of the location of the Jewish colonial plantations *per se*. In his brilliant study "'Still an Idea behind It': Trieste, Jewishness, and Zionism in 'Ulysses,'" Neil Davison extensively describes Joyce's exposure to Herzlian Zionism in Trieste and his growing interest in the latter. However, despite Davison's in-depth biographical research to support his argument, the textual evidence derived from Joyce's *Ulysses* remains indirect though not unpersuasive. Interestingly, in this study on Zionism and *Ulysses*, Davison specifically points to Joyce's 1918 notes for the episode of "Cyclops" with the hope of elaborating on the Zionist influence on Joyce's writing process. Although Zionism has never been explicitly mentioned in "Cyclops," its nationalist and political agenda does correspond with the Cyclopean themes of nationalism, antisemitism, and ethnic politics. In notesheet number eight for "Cyclops," Joyce puts down: "(Jesus did not Speak Hebrew)/120,000 settlers ... (Hebrew Gymnasium in Jaffa. 200?scholars)/Haifa polytechnical" (113.8–10). This entry bears explicit evidence of Joyce's Zionist source of information during his composition. With regard to the above lines from Joyce's notebook, Davison rightly signals their Zionist connection:

> The set alludes to the controversy over Hebrew as an ancient, liturgical language being reclaimed as modern speech and early Zionist efforts at Hebrew education in Palestine. Here, perhaps, Joyce is weighing the Gaelic Revival against Hebrew versions he witnessed in Trieste and uses them to inform the Citizen's denigration of linguistic-cultural nationalism in favor of aggressive republicanism [...].[24]

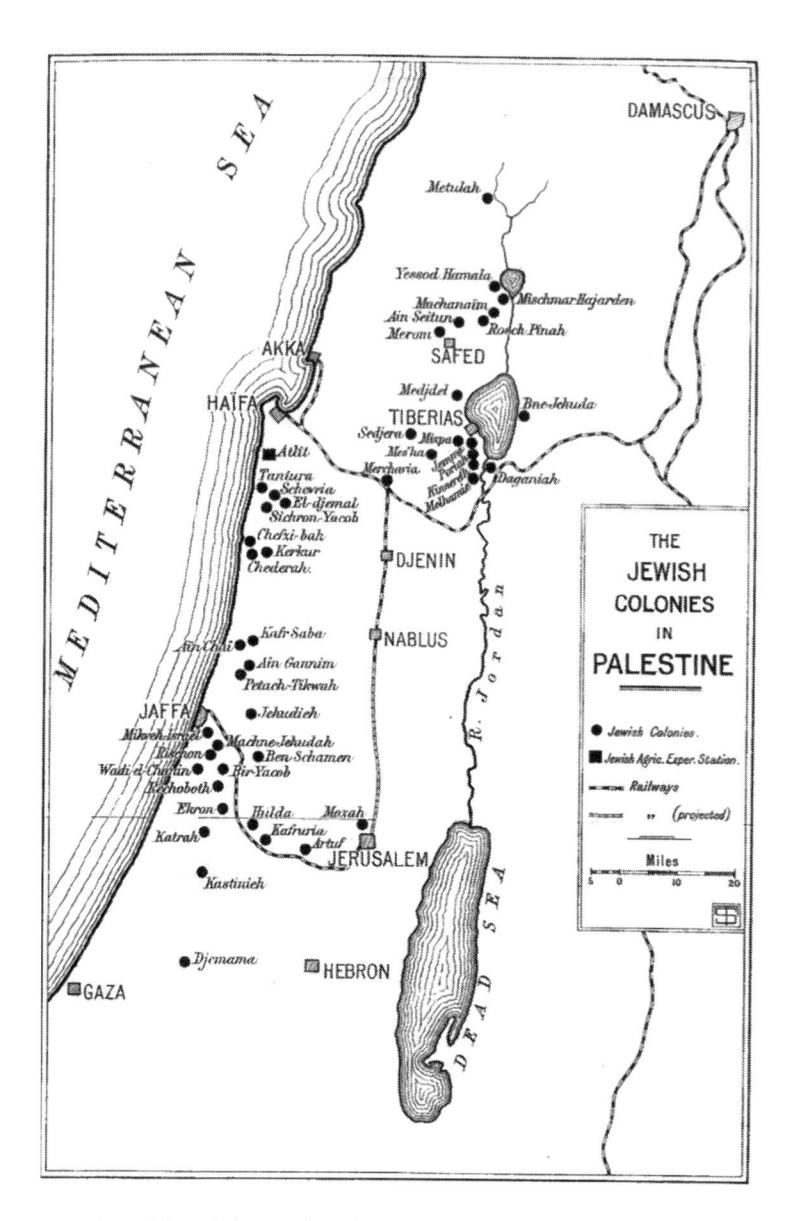

Fig. 2.1: Adopted from "The Jewish Colonies in Palestine" from *Zionism and the Jewish Future*, pp. 138–9.

Persuasive as Davison's argument sounds, it would be interesting to note that these Zionist allusions in fact very possibly come from the same source which Joyce had access to — *Zionism and the Jewish Future*. Dr. S. Brodetsky, another contributor to this collection, informs us of the foundation of a "Hebrew *Gymnasium* (Secondary School) in Jaffa" in 1906, which in his words was "another triumph of the Hebrew spirit in Palestine"[25]. As the pioneering movement in the revival of the Hebrew language in Palestine, it successfully carried out its "foundational idea [...] that all the teaching should be carried on in the national tongue"[26]. Following such pedagogical success in Jaffa, there emerged "the more ambitious project of establishing a Jewish Polytechnical Institute in Palestine"[27]. This is proposed with the aim to "complete the educational equipment of Palestinian Jewry" and to "provide the pupils of the secondary schools with the requisite higher technical education, so that they might become more useful workers in the newly established Jewish community"[28]. In "Cyclops" the debate about the Irish language revival uses similar discourse to Zionist pedagogy and the Hebrew language movement. The fact that throughout *Ulysses* Leopold Bloom anticipates and indeed foresees in "Circe" a new communal order of New Bloomusalem also echoes the Zionist appeal for a cooperative communal utopia in the Promised Land.

In the introduction to *Zionism and the Jewish Future*, Dr. Chaim Weizmann, later the first president of Israel (1949–52), clearly states the aim of the Zionist movement to be embedded in the return of the Jewish people to the Jewish land, especially that of Palestine. He claims:

> The task of Zionism [...] is to create a home for the Jewish people in Palestine, to make it possible for large numbers of Jews to settle there and live under conditions in which they can produce a type of life corresponding to the character and ideals of the Jewish people. When the aim of Zionism is accomplished, Palestine will be the home of the Jewish people, not because it contains all the Jews in the world, but because it will be the only place in the world where the Jews are masters of their own destiny, and the national centre to which all Jews will look as home and the source of all that is most essentially Jewish. *Palestine will be the country in which Jews are to be found, just as Ireland is the country in which Irishmen are to be found, though there are more Irishmen outside of Ireland than in it.*[29]

Bearing in mind the contemporary context of Irish nationalism at the time of writing, Weizmann's mention of the Irish and Irish diasporas here is particularly curious. Both Neil Davison and Brian Caraher indicate the prominent impact of the Balfour Declaration on Joyce's perspective toward Zionism as portrayed in *Ulysses*. In his groundbreaking study that significantly locates the 1917 Balfour Declaration alongside the Irish question within the historical context of post-Great-War Britain, Caraher singles out the colonial politics of the British support for the Zionist utopia — the establishment in Palestine of a national home for the Jewish people. He informs us that "Sir Ronald Storrs, the first Governor of British-mandated Palestine, welcomed the new arrangements, saying that they would facilitate 'forming for England a "little Jewish Ulster" in a sea of potentially hostile Arabism'"[30]. Storrs' "unsettling [parallel] of immediate post-Great War diplomatic and political attitudes in Britain to the past, present and future of Palestine and Ireland", Caraher observes, would in retrospect pave the way for the "utterly bizarre legacy [that] encompasses the strange sight of Palestinian and Irish 'tricolours' flying side by side" in contemporary West Belfast[31]. Such a cultural legacy of the Irish-Israel parallel, rather than simply "a colonial project of forceful resettlement in the Levant in pursuit of a lost homeland," he forcibly argues, would be the foundational appropriation at work in Joyce's propaganda of "Agendath Netaim: planters' company" in *Ulysses*[32].

In "Ithaca," Leopold Bloom remembers the wasteland-reclamation scheme advertised in the Agendath Netaim leaflet as his mind moves to possible financial sources for his ideal countryside estate Flowerville, which is mentioned just a few lines earlier. For him, the Agendath Netaim plantation plan would possibly be one of the "industrial channels" from which he could acquire "vast wealth" (*U* 17.1698) to provide for his personal consumerist desire. "Cultivation of orange plantations and melonfields[,] reafforestation," as well as "the utilization of waste paper, fells of sewer rodents, human excrement possessing chemical properties," are all outlined in his plans of possible industrial channels of income (*U* 17.1699–1703). In his contribution to *Zionism and the Jewish Future*, Tolkowsky details the economic industries at work in these Palestinian colonies: apart from orange and melon farming, he also mentions the broader scope of agricultural industry as well as the reafforesting of the land with eucalyptus trees for economic as well as environmental purposes. Other economic developments in these Palestinian Jewish colonies include sanitary facilities[33], a public transportation system including railways and other network systems, as well as the effective use of water resources by dam construction and the profitable manipulations of the rivers of the plain of Sharon and of Esdraelon[34]. These notions of public construction schemes in the Zionist colonies of Palestine curiously correspond thematically to many of Bloom's "schemes of wider scope" to follow (*U* 17.1709–42).

In his study, Caraher suggests an oblique association between "Baron Leopold Rothschild" (15.1848) and Baron Lionel Rothschild (1808–79), grandfather to the addressee of the 1917 Balfour Declaration. Interestingly, Tolkowsky introduces to us another Baron Rothschild who was closely associated with the Jewish Palestine plantation movement. Perhaps not coincidentally, Baron

Edmund de Rothschild, of Paris, became involved in the establishment of the Jewish colonies in Palestine out of his sympathy with the difficulties these colonies were encountering in the 1880s. He offered protection for four colonies "whose situation was most embarrassing" and later on, between 1884 and 1888, founded three more colonies[35]. Tolkowsky comments on the influence of Baron de Rothschild's contribution:

> ... whereas the essentially philanthropic system of colonization practised by Baron Edmund de Rothschild and the J.C.A. [Jewish Colonization Association] had only brought to Palestine immigrants who possessed little or no means, the expansion of the Zionist movement led to the influx into Palestine of a large number of middle-class Jews from all parts of the world, resolved to find in the country an outlet for their energies and for the small or moderate capital which they brought with them.[36]

Although his patronization of these Jewish colonies preceded the establishment of the Zionist organization in 1908, Baron de Rothschild's philanthropic empathy for these Palestinian Jewish establishments would have been of interest, to say the least, to Leopold Bloom.

Referring back to the 1917 Balfour Declaration, and *Zionism and the Jewish Culture* published a year earlier, we learn the explicit political Jewish-Irish linkage which would play such an essential role throughout *Ulysses,* not the least in the consistent appearances of Agendath Netaim from the fourth episode to the seventeenth one, in other words from soon after Leopold Bloom, our Irish Jew, entered the story, until his last episode "Ithaca," right before he recedes into the night. In such a way the Zionist idea of repossessing Palestine as the Jewish homeland becomes a point of reference throughout the book; indeed it maintains a material presence in Bloom's pocket throughout the day. With its Zionist propaganda "Agendath Netaim" has become, for the diasporic Jew Leopold Bloom, a promised-land vision of the Jewish utopia[37]. Nonetheless, it is substantialized as a utopian vision first made purchasable through consumerist propaganda, then rendered disposable, either initially by the Jewish butcher Moses Dlugacz or later in the day by Leopold Bloom as he puts it into flames to light his candles at home. At stake here is that the "Agendath Netaim" — the Zionist plantation in Palestine — is taken ironically as a piece of garbage to be recycled, or worse yet to be disposed of. Not only is it offered as a wrapping paper for the Jewish butcher, but it is also treated lightly in the case of Leopold Bloom, our Irish Jewish protagonist, as he picks up the leaflet randomly and eventually burns it. The shift of the leaflet's destiny from a token of a utopian (and in fact nationalistic) solution for contemporary political, economic and cultural turmoil to a piece of inflammable waste calls to question the interchangeability of wasteland

and Promised Land, and of waste and commodity. "Agendath Netaim" as such for Leopold Bloom in Dublin of 1904 is not so much of a practical solution to his contemporary and personal problems through monetary investment, neither is it for him an agreeable nationalistic scheme for reconciliation. Rather, portrayed as a utopian vision on the possibility of reclaiming wasteland, the leaflet of "Agendath Netaim" brings out a whole parallel dynamic of Jewish-Irish nationalist discourses, and in such a way Joyce also questions the (post/semi)colonial tension between industrial/infrastructural developments and nationalistic modernization. This is where waste and wasteland come into play not as distinctive substances of either utopia or debris, but as cultural artifacts of modernity in the face of nationalism and colonization.

II.

> *Surely, the State is the Sewer.*
> — Dominique Laporte, *History of Shit*, p. 56

Although Joyce may have used the historical context and the contemporary environmental agenda as part of his sources for his writings on sewage and wasteland, the historical references only underline Joyce's use of the material alongside his diverse narrative strands in *Ulysses*. Earlier in the episode of "Proteus," on Sandymount Strand around 11 a.m., Stephen Dedalus also had an encounter with waste and sewage:

> The grainy sand had gone from under his feet. His boot trod against a damp crackling mast, razorshells, squeaking pebbles, that on the unnumbered pebbles beats, wood sieved by the ship worm, lost Armada. Unwholesome sandflats waited to suck his treading soles, breathing upward sewage breath, a pocket of seaweed smouldered in seafire under a midden of man's ashes. He coasted them, walking warily. A porterbottle stood up, stogged to its waste, in the cakey sand dough. A sentinel: isle of dreadful thirst. Broken hoops on the shore; at the land a maze of dark cunning nets; farther away chalkscrawled backdoors and on the higher beach a dryingline with two crucified shirts. Ringsend: wigwams of brown steersmen and master mariners. Human shells. (*U* 3.147–57)

This is a condensed paragraph of the gathering of different "unwholesome" relics from literature, history, civilization, (alcoholic) consumption, and even life remains. Stephen notices the "unwholesome sandflats [waiting] to suck his treading soles [while] breathing upward sewage breath ..." (*U* 3.150–51) and cautiously makes a detour in fear of getting his soles contaminated. With regards to this line, Gifford's annotation provides us with a glimpse of contemporary sewage problems which explain Joyce's further elaboration of this modern civil problem in later chapters:

> Much of Dublin's sewage was emptied untreated into the Liffey and its tributaries, so the streams in the metropolis were little better than open sewers, and the inshore waters of Dublin Bay, particularly just south of the mouth of the Liffey, where Stephen is walking, were notoriously polluted.[38]

Stephen's consciousness of the state of sewage pollution on Sandymount Strand, alongside his awareness of the relics of history, literature, and civilization, together highlight the binary conflicts between civilization and waste, imperialism and primitivism, power and shame, and consumption and death. "[Sewage']s unwelcome appearance on the shoreline or gushing out of storm water channels straight into the ocean," Hawkins observes, "triggers all sorts of anxieties: possible epidemics, a site of pleasure and hedonism threatened, fear of a poisoned world"[39]. By making an intentional detour, Stephen for one is evidently aware of the threats associated with pollution by the dispossessed of modern civilization — although his anxiety seems to reside more on a personal level than on an environmental one. In the meantime, he is also aware of the overwhelming power of the whirlpool of desperation and frustration from the relics of the history of this "isle of dreadful thirst" (U 3.153–4). Just as history is a nightmare from which Stephen is trying to awake (U 2.377), his anxiety about history is paralleled with his anxiety about the sewage-polluted sand on Sandymount Strand.

Stephen's horror about the waste on Sandymount beach reveals to us not so much his environmental concern than his fear of death and his obsession with the cyclical concept of history. Contemplations of death linger in his thoughts as he strides along the beach; as he notices the distasteful relics of history and civilization, the thought of "human shells" (U 3.157) also flashes through his mind. Dominique Laporte curiously indicates that "[c]orpses are no more and no less than waste that one buries"[40], and that "the Christian West has long responded with equal terror to the smell of shit and of corpses" as "one finds significant parallels in the morbid effects attributed to their respective odors as well as in the desire to hold them both at bay"[41]. Apart from their shared attribute as relics of society, excrement and corpses are also both deemed by Laporte to be colonial objects that reflect the imperial structure of power and oppression. For Laporte, shit — as well as a corpse — arouses fear in men and stirs up anxiety as a result of the imbalance of power between the imperial and the colonized subjects. Using a white-versus-black racial analogy, Laporte writes:

> To the black man, the white man looks and smells like a corpse. To the white man, the black man has the color and odor of shit. Their mutual hatred is based on a reciprocal recognition: the white man hates the black man for exposing that masked and hidden part of himself. The black man hates the white man's need to pull himself up from the earth. (The conqueror pulls himself from his native soil to till the soil of another, to exploit its capacity for production and, in so doing, cultivates it and cleanses its inhabitants.) The black man sees in the white man's need the blind arrogance of one who thinks himself immortal. But he who brings civilization cannot help but feel immortal. This is why he smells like a corpse: he is constituted by the return of that repressed "remnant of earth," which clings to him as much as to any man.[42]

Laporte's argument offers an alternative reading of the politics of waste in view of the colonial question. In the context of *Ulysses*, just as Bloom as a diasporic Irish Jew associates human excretion with the reclamation of the (Palestinian/Irish) wasteland, Stephen as the Irish colonial object sees waste as the feces of the cyclical movement of history. Towards the end of the third episode, Stephen's thought wanders back to a corpse, death, excrement and the life circle:

> Bag of corpsegas sopping in soul brine. A quiver of minnows, fat of a spongy titbit, flash through the slits of his buttoned trouserfly. God becomes man becomes fish becomes barnacle goose becomes featherbed mountain. Dead breaths I living breathe, tread dead dust, devour a urinous offal from all dead. Hauled stark over the gunwale he breathes upward the stench of his green grave, his leprous nosehole snoring to the sun. (*U* 3.476–81)

In his book *On Garbage* John Scanlan indicates that traces of material disorder like dirt, filth, and dust "become symbolic of garbage not simply because they represent displaced matter, but more precisely because of 'things' such as their formlessness (i.e., they may have been something once but are now nothing)"[43]. Whereas "Proteus" in Greek mythology is renowned for his power to transform in *The Odyssey*, Stephen's contemplation on life transformations in the ecosystem seems to refer us back to the Viconian model of the cyclical structure of history[44]. However, it was not simply the formlessness of waste itself but also the transformative attributes of waste as well as lives and materials with which Stephen engages his thinking. Waste (or a corpse) is more than simply the opposite of life or civilization; it is part of the cyclical process of transformation — be it the ecosystem or cyclical history — in which humans, animals, and industrial products all take part.

Stephen's uncompromising indifference toward waste matters is, Michael Rubinstein contends in his book *Public Works*, derived from his distaste toward public works in general[45]. This is what Rubenstein terms as "the antinomy between utility and the aesthetic in the Irish imagination"[46], which finds its full expression in the contrast between Stephen Dedalus the artist and Leopold Bloom the down-to-earth man, and their attitudes toward sewage matters. Contrary to Stephen's impassiveness, Bloom's thoughts on public works linger around "[t]he reclamation of dunams of waste arenary soil" as well as "[t]e utilisation of waste paper, fells of sewer rodents, human excrement possessing chemical properties" (*U* 17.1701–3), and then even further unto "schemes of wider scope" (*U* 1709–43). The Stephen-Bloom contrast is not the only means through which Joyce illustrates such conflict between art and utility in a period of history in which modern Ireland was coming of age. Another vivid example of such antinomy can be found in Professor MacHugh's joke in "Aeolus."

In "Aeolus," in the *Freeman's Journal* office, the civic "solution" to the modern anxiety for human excrement is brought up by Professor MacHugh in his brief lecture on imperialism and civilization:

> **THE GRANDEUR THAT WAS ROME**
> – Wait a moment, professor MacHugh said, raising two quiet claws. We mustn't be led away by words, by sounds of words. We think of Rome, imperial, imperious, imperative.
> He extended elocutionary arms from frayed stained shirtcuffs, pausing:
> – What was their civilisation? Vast, I allow: but vile. *Cloacae*: sewers. The jews in the wilderness and on the mountaintop said: *It is meet to be here. Let us build an altar to Jehovah*. The Roman, like the Englishman who follows in his footsteps, brought to every new shore on which he set his foot (on our shore he never set it) only his cloacal obsession. He gazed about him in his toga and he said: *It is meet to be here. Let us construct a watercloset*. (*U* 7.483–95)

Immediately after Professor MacHugh's remarks on the construction of lavatories, Lenehan directs the focus of their conversation to the Irish:

> – Which they accordingly did do, Lenehan said. Our old ancient ancestors, as we read in the first chapter of Guinness's, were partial to the running stream.
> – They were nature's gentlemen, J. J. O'Molloy murmured. But we have also Roman law.
> – And Pontius Pilate is its prophet, professor MacHugh responded. (*U* 7.496–501)

Here again, we see another distinct contrast using the metaphor of sewers, yet on a slightly different scale. The thrust of the joke points toward British empire building by ridiculing the practicality of the Anglo-Saxons in juxtaposition with the characteristics of the Irish — poetic, passionate, impractical. But conversely, the Irish resistance to the pragmatics of planning and building also serves to justify the annexation of Ireland under the Act of Union of Great Britain and Ireland in 1800: as Matthew Arnold observes, "As in material civilisation he has been ineffectual, so has the Celt been ineffectual in politics"[47]. The stereotypical difference between the Saxon and the Celt in MacHugh's joke turns on "the distinction between utility and the aesthetic"[48], and, "along with the history of utilitarian projects and modernising efforts in Ireland, reveals a stereotypical Irish attitude toward utilitarianism and modernisation"[49].

To uncover this Irish complex concerning public utility and modernization, Rubinstein directs us to the Great Famine of the nineteenth century. David Lloyd writes that the Famine was a "colonial catastrophe" brought about in the name of modernization through "the deliberate use of famine relief projects, eviction, and emigration under duress to eradicate ways of life that had been recalcitrant to capitalism"[50]. The Irish Board of Works was founded in 1831 to undertake relatively useful projects including some successful drainage schemes. But during the Great Famine, the board went on to take on all sorts of relief schemes "that were to become well-known emblems of and the monuments to the mistreatment of the Irish at British hands"[51]. This was, as Rubenstein points out, "the era of the famous 'famine roads' that 'began nowhere and ended nowhere,' of 'walls round nothing' and 'piers where no boats could land'"[52]. Acting on strict orders that famine relief be contingent, the board was enforcing work on unnecessary public utilities when there was no other work to be done. This episode in Irish history marked a "complete emptying out of the concept of utility, insisting that relief takes the form of utility without the function, giving rise to the surreal existence of previously unimaginable human-made artefacts: an ornamental road, for example"[53]. To make matters worse, the low wage earned by individual involvement in infrastructural work could not have been earned otherwise — hence prohibiting the Irish from farming their own lands and contributing to the vicious circle that exacerbated the effects of the Great Famine, and, in such a way, aggravated the Irish hostility against public utility. "If not deliberately genocidal," David Lloyd comments, "British policy nonetheless sought to make of the Famine a means both to the clearance of what they regarded as a 'redundant' population and to the transformation and modernisation of Irish agricultural processes"[54]. The Famine was, Joe Cleary notes, an example of colonial modernization, "a drastic reduction of population to market rather than expansion of market to population," and an imperial act of "political modernization" that "meant [for the Irish] a diminishment rather than an extension of political sovereignty"[55]. Indeed, these attributes — the compression of the processes of modernization from "several centuries" to just several short years, and the diminishment of political sovereignty — mark the difference in the underdeveloped identity between the Irish colonial experience of modernization and the European experience. The Irish resistance to infrastructural works was hence formed as the inevitable result of their colonial experience of modernization from the Great Famine onwards.

While the Irish "antinomy of utility and the aesthetic" through "the twin legacies of colonial modernisation and the racialized discourse of the poetic Celt"[56] is ruthlessly addressed by Joyce by means of Professor MacHugh's joke on Roman watercloset, the passage further delves into the question of civilisation and the power politics of the sewer. Lenehan's direct response, "Our old ancient ancestors, as we read in the first chapter of Guinness's, were partial to the running stream" (*U* 7.496–8), further emphasizes the contrast between nature and civilization in the context of the water closet. Regarding Professor MacHugh's reference to England, Laporte informs us of the emerging attention to cleanliness within Victorian England:

> Given the rage for hygiene that swept through nineteenth-century Europe, and England's role in promoting an extremely tasteful range of products that catered to new notions of cleanliness, order, and, by extension, beauty, the observation of Professor MacHugh and his acolytes could not be more *a propos*. In France, as elsewhere, a new kind of fixture would forever be attributed to its country of origin, and by the end of the nineteenth century, one only heard talk of "English basins" and "English urinals."[57]

Bearing in mind the eminence of English water closets and the Empire's attention to private and public hygienic developments, it is crucial that we as readers of *Ulysses* also take into account the interwoven narratives of imperialism and nationalism alongside descriptions of civil and domestic sewage facilities. Indeed Joyce's metaphorical use of the parallel between the Roman Empire and Irish natural law in the case of toilet facilities interestingly associates the intriguing relationship between civic order and private sewage with consumption and imperial bureaucracy. As Gifford annotates, Lenehan's triadic pun (*U* 7.496–8) implies that: "as the Jews were to their altars, and the Romans to their waterclosets, so the Irish are to their drink"[58]. This triadic reference does not simply indicate the religious, political, and alcoholic influences on the Irish (as exemplified by Stephen Dedalus and how these three trigger his thoughts throughout *Ulysses*), but also starkly points to the focal interest in each culture. Just as building altars is central to the Israelites in the Book of Genesis in the *Old Testament*, so is building water closets, as Joyce possibly implies here, an act of worship and sacrifice to the Roman order of civilization, and in the case of the Irish, the culture of drinking. On this Laporte further states that:

> The imperialist era effected a partial revival of the Roman deification of human secretions, albeit in the guise of an emergent atheism. Rome, it is true, demonstrated an even greater attachment to all those things the public privy stood for in the Victorian

West. But the nineteenth century's compulsive cleanliness cannot adequately explain the architectural abandon with which its sites of shit were seized: these were places of commemoration. They appear almost as shrines, where civilized man deposited offerings and prayers to ward off the very awareness of his primordial origins that the revival of antique customs paradoxically invoked.[59]

In this sense, private and public sewage systems provide covering for modern men to hide their excrement as well as their acts of excretion. By concealing waste from the sight of the public, modern cities have enabled a modern system of classification and separation. The birth of the modern sewage system along with the rise of modern urbanization not only significantly transforms the environmental perspective toward modern cities, but also shifts the human relationship with the environment as well as with their own bodies. The human awareness of their own excrement intensifies as the modern experience of separation from human waste becomes a common practice as a result of the demand for cleanliness and order. Such enforced separation, by way of concealment, aggravates the alienation of modern men. As demands for order and hygiene become disciplines to enforce hierarchical ideologies and power in a modern city, the sense of shame and oppression of the human subject also become one of the problems of modern citizens. Stephen Dedalus's encounter with the sewage-polluted sand on Sandymount beach confronts with the reality of human waste environmentally as well as historically. Toward the end of the day, as he leads Bloom to urinate in the open darkness (*U* 17.1185–98), his urination is more than another simple act of relieving oneself — it is an act of rebellion against the imperial system of sewage, of power and the inquisitional gaze of the empire. Commenting on Joyce's cloacal obsession, Gifford informs us that in *Finnegans Wake* Joyce "explicitly associates micturition with poetic creativity and with the writing of *Ulysses*"[60]. If, as Laporte brilliantly puts, "the State is the Sewer [...] not just because it spews divine law from its ravenous mouth, but because it reigns as the law of cleanliness above its sewers"[61], then the micturition of Stephen and Bloom in the open-air in the dark (*U* 17.1185–98) would arguably point to Joyce's poetic composition of *Ulysses* as a creative outlet to relieve himself of the surveillance of the Sewage system — the State.

III.

The sewer is the conscience of the city.

— Victor Hugo, *Les Miserables*

Like Swift and another living Irish writer [G. B. Shaw], Mr. Joyce has a cloacal obsession. He would bring back into the general picture of life aspects which modern

> *drainage and modern decorum have taken out of ordinary intercourse and conversation.*
>
> – H. G. Wells, "James Joyce," *New Republic* 10 (10 March 1917), p. 159

Thornton suggests that H. G. Wells's criticism of *A Portrait of the Artist as a Young Man* may have had a direct influence on Joyce's composition of Professor MacHugh's mentioning of the Roman's (and Englishman's) "cloacal obsession"[62]:

> He [Professor MacHugh] extended elocutionary arms from frayed stained shirtcuffs, pausing:
> – What was their civilisation? Vast, I allow: but vile. *Cloacae*: sewers. The jews in the wilderness and on the mountaintop said: *It is meet to be here. Let us build an altar to Jehovah*. The Roman, like the Englishman who follows in his footsteps, brought to every new shore on which he set his foot (on our shore he never set it) only his cloacal obsession. He gazed about him in his toga and he said: *It is meet to be here. Let us construct a watercloset.*
> – Which they accordingly did do, Lenehan said. Our old ancient ancestors, as we read in the first chapter of Guinness's, were partial to the running stream. (*U* 7.487–98)

"Aeolus" was first published in serial form in *The Little Review* in October 1918, and according to Gabler's synoptic edition of *Ulysses*, the mention of "cloacal obsession" was one of Joyce's initial designs for this passage. It is hence crucial to consider the question of sewage and relevant facilities in the context of H. G. Wells's critical review of *Portrait*. If, as I have previously pointed out, Joyce indeed makes the sewage-versus-state metaphor a referential response to H. G. Wells's critique of his work, such association with the imperial discourse of hygienics puts the text into a challenging position for the assuming voice of Anglo-Irish cultural nationalism, a tradition H. G. Wells shares. Andrew Gibson says that "as one of the most prominent in a chorus of similar English and Anglo-Irish voices raised early in the 1920s, … [H.G.] Wells effectively identified the multifaceted outrageousness of *Ulysses* with a political and cultural offensive"[63]. These voices, as Gibson observes, tended to express similar opinions: that "Joyce's works might be 'powerful', but they were also shocking in their 'squalor', 'dirtiness', 'unpleasantness', 'coarseness', and 'astounding bad manners'"[64]. As Gibson comprehensively argues throughout *Joyce's Revenge*, the book *Ulysses* was written as a witness to the process of conflict and conciliation among the Anglo-Irish, the English and the Irish in the period 1880–1920. Joyce's "coarse" writing of the dirtiness, the excrement, and the sewage throughout *Ulysses* also seems to undermine the dominant discourses of imperial order and Anglo-Irish cultural nationalism. H. G. Wells's comment hence enhances Laporte's "the State is Sewer" argument in the context of waste and power, and makes it clear that Joyce's "cloacal obsession" is in fact not simply a vulgar attack on the cultured discourses of modernity, but more significantly an alternative discourse of mundane modernity — one which has been taken out of circulation by modern drainage and modern decorum, as H. G. Wells explicitly states, under the imperial structure of discipline.

Just as Gay Hawkins makes it clear that what we want to get rid of not only "tells us who we are" but also "*makes* us who we are"[65], in the same way, without examining dispossessed waste in *Ulysses,* the city of fin-de-siècle Dublin cannot be fully captured and identified. In fact, waste, sewage and the odor of asphalt were so common to urban experiences of Dublin that it is impossible to neglect their existence. As Cheryl Herr's interesting research addresses, Dublin — historically a midden town — is metonymically referred to as "an asphalt, a garbage dump in which many bottles and many more corks have found their way"[66]. Pointing it out to his potential publisher for *Dubliners,* Joyce himself acknowledged that "people might be willing to pay for the special odour of corruption which, I hope, floats over my stories"[67]. Later in another letter to Grant Richards in June of 1906, he defends himself regarding the same issue: "It is not my fault that the odour of asphalt and old weeds and offal hangs round my stories"[68]. However, from the composition of *Dubliners, A Portrait of the Artist as a Young Man,* to that of *Ulysses,* Joyce's attitude toward and his handling of scatological matters went through noticeable changes[69]: from the overall passive, and even resistant (exemplified in Stephen Dedalus), stance toward public utilities concerning waste and sewage matters throughout *Dubliners* and *Portrait, Ulysses* takes on such a diversity of perspectives (Stephen's and Bloom's being two of the many voices) through Joyce's use of free indirect discourse(s) that readers are provided with a plethora of voices, some more outspoken than the others, regarding the modern question of sewage. One of those distinct voices, curiously, comes from the sewage itself:

> From its sluice in Wood quay wall under Tom Devan's office Poddle river hung out in fealty a tongue of liquid sewage. (*U* 10.1196–7)

It is probably not surprising that the sewage was given such a presence among the saluting and unsaluting Dubliners in the presence of the viceregal cavalcade in the "Wandering Rocks" episode, in which "the mind of the city [...] is both mechanical and ironic"[70]. Interestingly, as Clive Hart points out, the Poddle empties into the Liffey not at Wood Quay but Wellington Quay. However, Joyce's change of details allows the sewage to emerge ironically from just beneath the building of the Dublin Corporation Cleansing Department Offices, which were located at 15–16 Wood Quay[71]. Another fact about the Poddle, significantly, was that it was the source of much of Dublin's drinking water before the

Dublin Corporation Waterworks Act of 1861[72]. That such a source of potable water would become, by 1904, "a tongue of liquid sewage" beneath the institution whose name was dedicated to its maintenance, The Dublin Corporation Cleansing Department, reflects a tragicomic illustration of urban ecological degradation and the futility of infrastructural modernisation under colonisation, or, in Joe Cleary's term, "colonial modernization"[73].

On Joyce's referential use of the Poddle and Wood Quay, Rubenstein's study further informs us:

> Wood Quay was the site of the very first Viking settlement in Dubin in 841 AD, just east of the point where the Poddle met the Liffey. Those early Norse settlers docked their ships in a body of stagnant water at the river's intersection that they called "Black Pool," a phrase taken from the Irish "Dubh Linn," which was borrowed from the Irish Christian monastery they conquered when they settled there.[74]

Here we are reminded of Victor Hugo's remark on the sewer being the conscience of the city, as I quoted in the epigraph of this section. Indeed, for Joyce to relocate the river's confluence in time as in space is to give the river a "sewer of consciousness" whose history traces back to the very first colonial settlement in the "Black Pool" which has, ironically, metempsychosed into a black pool of liquid sewage. Starting from the Irish word "Dubh Linn," its story is the city's whole history of conquest, by the Celts, the Vikings, the Normans, and, of course, the latest viceregal cavalcade and the British imperial rule it represents. Through the city's "sewer of consciousness" and the sewer's mocking posture toward the cavalcade, Joyce grants Dublin an alternative voice by a subjectless subjectivity which, "descended from the utilitarian ideal of governmentality, and as a ubiquitous urban infrastructure it advertently offers a point of view that is no longer rooted in human subjectivity"[75]. In such a way, liquid sewage serves as a kind of subjectless subjectivity among the many voices and perspectives that contribute to the "mind of the city"[76] of Dublin, an organism made possible only through the imagined urban community with streams and sewers of consciousnesses.

From the "sewer of consciousness" of the city, let us now move onward to consider the way Joyce constructs an alternative ecosystem according to waste matters. Hawkins reminds us that waste matter is beyond what is simply deemed as a result of material usage or disposal — it is rather the modern product of selection under the ideological system of classification and consumption:

> As much as putting out the garbage may feel like one of the most ordinary and tedious aspects of everyday life, it is a cultural performance, an organized sequence of material practices that deploys certain technologies, bodily techniques, and assumptions. And in this performance waste matter is both defined and removed; a sense of order is

established and a particular subject is made. *Waste, then, isn't a fixed category of things; it is an effect of classification and relations.*[77]

This connection between commodification and subjectivity is not unknown to Joycean readers. In his *Ulysses, Capitalism, and Colonialism*, M. Keith Booker indicates: "For individual subjects, the most crippling effect of the processes of reification and commodification in a capitalist economy involves the reduction of human beings and the relationships among them to the status of objects"[78]. Just as the capitalist system of modernity objectifies civic connections as well as individual subjectivity in such a way that human relationships have become dysfunctional[79], the agenda of recycling and waste reclamation, on the other hand, seems to provide a reversed system of classification and selectivity for modern society. With Joyce's emphasis on the recycled or reclassified "waste" exemplified in Bloom's conceivable waste reclamation scheme (*U* 17.1699–1708), liquid sewage is given a narrative consciousness (*U* 10.1196-7), and even a piece of disposable wrapping paper that awakens a whole new dynamic to Bloom's imagination on the Zionist Promised Land. In such a way, Joyce not only offers an alternative cultural politics of the commodity in capitalist society but also signals a redemptive politics for the problem of alienation in the modern fin-de-siècle society.

It is hence curious yet not surprising that the episode of "Lestrygonians," the peristaltic chapter aimed at revealing the cannibalism of the capitalist system of consumption, opens with a list of edible commodities as well as the introduction of a throwaway:

> A sombre Y. M. C. A. young man, watchful among the warm sweet fumes of Graham Lemon's, placed a throwaway in the hand of Mr Bloom. (*U* 8.5-6)

Leopold Bloom is handed an evangelical Y. M. C. A. leaflet which contains the notice "Dr John Alexander Dowie restorer of the church of Zion" is coming to Ireland — "Elijah is coming" (*U* 8.13-14). On the one hand, this disposable piece of religious propaganda contributes to the narrative of the commodity in this chapter, connecting the throwaway to the restoration of Zion. It is worth noting that in the meanwhile the Jewish wasteland reclamation project hasn't completely departed from Bloom's mind: the Agendath Netaim leaflet still lies in his pocket, and it will be brought up again towards the end of this episode as he reaches "[h]is hasty hand" into his pocket and "took out [...] read unfolded Agendath Netaim" (*U* 8.1183-4) in panic to avoid the embarrassment of direct contact with Boylan. Later on, this piece of "Elijah is coming" throwaway will be involved in the diverse urban wandering routes through Dublin in the episode of "Wandering Rocks":

> Elijah, skiff, light crumpled throwaway, sailed eastward by flanks of ships and trawlers, amid an archipelago of corks, beyond new Wapping street past Benson's ferry, and by the threemasted schooner *Rosevean* from Bridgwater with bricks. (*U* 10.1096-9)

In this way, the throwaway marks its own civic route along the River Liffey, probably the same route as sewage, down towards the Irish Sea, possibly even toward Sandymount Strand. It joins the Poddle's "sewer of consciousness" in constituting another often-overlooked subaltern voice in harmony with the other human storylines of the episode — and indeed of the entire book.

In "Lestrygonians," as the hungry Bloom consumes his light lunch and the glass of burgundy at Davy Byrne's, his thoughts again wander off:

> .... Orangegroves for instance. Need artificial irrigation. Bleibtreustrasse. Yes but what about oysters. Unsightly like a clot of phlegm. Filthy shells. Devil to open them too. Who found them out? Garbage, sewage they feed on. Fizz and Red bank oysters. Effect on the sexual. (*U* 8.862-6)

It is worth noting here that Joyce again associates the Agendath Netaim scheme (as the words "orangegroves" and "Bleibtreustrasse" indicate) with consumption and waste. But instead of simply limiting it all to the idea of consumption as the exchange of commodities, this time he brings the consumption of food into the picture. Radical Zionist propaganda, garbage, food, and sex — these few lines are, for Joyce's contemporary readers, but one intense example of what Hawkins terms "the politics of disturbance"[80]:

> In experiences of horror and abjection we confront the constitutive uncertainty of naturalized cultural distinctions. There is no doubt that shit (like sex) can be very disturbing, and this is why it is of inestimable value for understanding the contingency of cultural boundaries.[81]

Just as disturbing as the mention of waste and sex is probably the interchangeability of waste, food, and sex as Joyce points out here. With his "cloacal obsession" Joyce does not simply undermine the hierarchical structure of power in the social construction of order and cleanliness; he also introduces a system of interchangeable relations which disrupts the binary divisions of private/public and primitive/ civilized, making it a fluid interactive network instead. Hawkins says that "sewers link us to the state without any sense of direct intervention" in such a way that they are literally "where citizenship and subjectivity intermingle"[82]. We see this interaction at work from Leopold Bloom's private time in the water closet (*U* 4.500–517) to his potential wealth-gaining plan through waste resources (*U* 17.1701–08). From the privatization of excretion to the publicised civil propaganda of waste recycling, Joyce illustrates that the modern citizen's relationship with waste also reflects his position within social powers structure under the surveillance of the state.

From the perspective of the state, "shit [has] to be excluded from the official in order to establish the state's power to cleanse" — hence "the focused privatization of excrement [is] a way of expressing state power"[83]. Shit therefore becomes "a political object through the process of making it individual or private responsibility, making its producers legal proprietors"[84]. In this way, "property and propriety become linked through the politics of shit"[85]. From Bloom's private time in his very own water closet to his industrial waste regeneration project for Ireland, substantial personal excrement would be conceivably be transformed into a collective State possession not only in its purpose but also in its ownership.

Whereas human excrement and other domestic waste flowing into the drainage system become possessions of the State, those separately and intentionally disposed of in private gardens would, in turn, constitute an alternative narrative of garden politics, liberation and anarchism[86]. We are reminded again of how Bloom pays attention to the interrelationship between sewage and gardening on his way to the water closet:

> He went out through the backdoor into the garden [...]. He bent down to regard a lean file of spearmint growing by the wall. Make a summerhouse here. Scarlet runners. Virginia creepers. Want to manure the whole place over, scabby soil. A coat of liver of sulphur. All soil like that without dung. Household slops. Loam, what is this that is? The hens in the next garden: their droppings are very good top dressing. Best of all though are the cattle, especially when they are fed on those oilcakes. Mulch of dung. [...]. (U 4.472–81)

As I've discussed at length in the previous chapter, Bloom's attentiveness to gardening would later be incorporated to his ideal self-portrait as Bloom of Flowerville (U 17.1551–87). Here in "Calypso" en route to relieving himself, his interior monologue directs our attention to his garden not from the botanical point of view but instead from the perspective of organic gardening. Dung, household slops, hen droppings and especially cattle dung in the edible garden occupy equal if not more space than the edible vegetations in the garden for Bloom. A similar "cloacal obsession" and focus on consumption would recur in "Lestrygonians" as Bloom ponders the garbage/sewage-feeding oysters (U 8.865). If the sewage system operates as a state-dominated structure demanding order while gaining power from the transfer of possessions, organic gardening proposes an alternative, and indeed anarchistic ecosystem that is not hierarchical but fluid and interchangeable. Hawkins notices "in composting campaigns organic waste is often represented as beautiful, its formless disorder reconstituted as an aesthetic of abundance; scraps, discards, leftovers becoming soil"[87]. In this sense, organicism bears a resemblance to the utopian community gardens set up by socialist and liberalist idealists to bring about a transformation of society through a vibrant grassroots community — which vision our very own Leopold Bloom might likely be sharing.

Joyce's explicit association of "micturition with poetic creativity and with the writing of *Ulysses*"[88] reminds readers of another fact about this very scene: Bloom is reading a contemporary popular journal *Tibits*[89] while defecating in the watercloset located in his garden (*U* 4.500–540). It is of specific interest that the popular journal Bloom reads in the water closet ends up serving as his toilet roll as "he t[ears] away half the prize story sharply and wipes himself with it" (*U* 4.537). Here is another graphic exchange between creative writing and human excrement. However, in contrast with the freedom of urinating in the open-air and in the dark (as we find Stephen and Bloom later in the day in the episode of "Ithaca"), here we witness the gesture of reading alongside the act of defecating, with the creative piece of writing ending up as toilet paper and excrement in the sewage system. As Bloom contemplates the sewage-fed oysters' "effect on the sexual" (*U* 8.862-6), the interchangeability of waste, food, and sex is also at work through popular journalism, whose readers feed on popular pornographic writing for its effect on the sexual. In a way, Joyce seems to be ridiculing the so-called popular journalism that caters to the public taste in a way similar to human excretion going into the drainage system under the surveillance of the State. In another way, by underscoring the sewage system and its contribution to modern (eco)systems and infrastructure, he seems to indicate that popular journalism and creative writing are also part of this ecosystem of consumption and power that inhibits the modern city of Dublin.

IV.

Of "Lestrygonians" episode Declan Kiberd writes: "Joyce recognizes that the processes of life run through decay, death and transformation: if newspaper can become food-holders, then waste really can be turned to use"[90]. While Joyce makes newspaper food-holders, cut ad-sheets sausage-wrappers, and an evangelical leaflet a pedestrian presence, he also makes the transformation from one-way "metempsychosis" (*U* 4.375) (a human being "changed into an animal or a tree" (*U* 4.376)) into interchangeable relations among human, commodities, and even disposables. Whereas Stephen Dedalus thinks more of linear eco-transformation: "God becomes man becomes fish becomes barnacle goose becomes featherbed mountain" (*U* 3.477–79), Leopold Bloom's view of the ecosystem involves humans, animals and waste in a system of consumption and bureaucratic order. While Stephen's rather historical, poetical approach to the ecosystem seems to ignore (to the point of resistance) modern commodities and waste, Bloom's ecosystem embraces commodities and waste in such a way that they are indeed an indispensable element of the cycle. Be it ecosystem of organicism, the capitalist system of profitability, or modern narratives about obsolete objects, Joyce incorporates waste in his narrative in a way that challenges the binary classification/division between human and waste, civilization and garbage.

From the reclamation of the wasteland (through the hidden discourses of Agendath Netaim and Bloom's industrial scheme (*U* 17.1699–1708)) to the reclamation of waste (through excrement, garbage, and sewage, as previously discussed in this chapter), in *Ulysses* waste and wasteland have been transformed from debris to cultural artefacts of modernity in the face of colonial modernism and consumerism. Yet Joyce's "cloacal obsession" doesn't stop there. By "bring[ing] back into the general picture aspects which modern drainage and modern decorum have taken out of ordinary intercourse and conversation"[91], he also brings back to the "ordinary discourses" of Dublin the scatological consciousness of the imagined community of waste matters. With the "sewer of consciousness" infiltrating the modern city of Dublin, Joyce's alternative scatological narrative allows an almost subaltern subjectless subjectivity to engage with the city's diverse voices on colonial modernization and modernity. And whereas the Irish subject relates to the sewage system through state sovereignty and the antinomy between utilitarianism and aesthetics, Joyce's rendering of the "sewer of consciousness," even one with a cynical attitude toward the cavalcade (its tongue hung out in tribute at the cavalcade's passing), seems to provide the sewer with a redemptively anticolonial existence in its contribution to the civic discourses of the city. Just as "the utility of waste is [...] a revival"[92], so is the redeemed waste itself in its newly given subjective consciousness. For Joyce, of course, waste is not simply part of the modern sewage system — it has to be read in the context of edible and poetic consumption, as well as part of a fluid ecosystem of interrelationships and inter-discourses.

Concerning Joyce's attitude toward infrastructure, Rubenstein boldly claims: "Public works [...] are Joyce's answer to the depredations of modernity, as well as his affirmation of modernity's utopian promise"[93]. Although I am not fully convinced by Rubenstein's argument, Joyce's concern for public works and modern environmental problems of sewage is clearly reflected in his writings, *Ulysses* in particular. Instead of advocating infrastructures, or the environment *per se*, Joyce allows these peripheral subjectless subjectivities to voice for themselves. Whether with a contemptuous gesture of salute toward the imperial parade (at Wood Quay) or in industrial details on outlining infrastructural schemes (*U* 17.1698-1743), these "sewers of consciousness" have indeed become the answers to their own questions on modernity. With the sewer's urban narratives conjoined with other diverse discourses of the city and its citizens, maybe I can boldly say *Ulysses* is, after all, an ecological text, as well as a modernist and postcolonial one? In the following two chapters (Part II), we'll spend more time considering the question of nature, nation, and ecosystem in *Ulysses,* with particular emphasis on the chapters of "Aeolus" and "Cyclops."

Notes

1. See Don Gifford's annotation 4.191-92, p. 74.
2. P. Anderson Graham, *Reclaiming the Waste: Britain's Most Urgent Problem* (London: Country Life Library, 1916), p. 1.
3. See Cheryl Herr's "Joyce and the Everynight" in Eco-Joyce: The Environmental Imagination of James Joyce, pp. 38–58, and Michael Rubenstein's Public Works: Infrastructure, Irish Modernism, and the Postcolonial. Notre Dame: University of Notre Dame Press, 2010.
4. Rubenstein, p. 70.
5. H.G. Wells, "James Joyce", *New Republic*, 10 March 1917.
6. Gillian Whiteley, *Junk: Art and Politics of Trash* (London: I.B. Tauris & Co., 2011), p. 16.
7. Ibid., p. 24.
8. Gay Hawkins, *The Ethics of Waste: How We Relate to Rubbish* (Rowman & Littlefield, 2006), p. 2.
9. Ibid., p. 2.
10. Shoar and Stam, pp. 42–3, qtd. in Whiteley 7, Italics mine.
11. Whiteley, p. 25.
12. Ibid.
13. Hawkins, p. 4.
14. I'll talk more on sewage in a later section of this chapter.
15. For more detailed discussion on this issue of afforestation in Ireland, see the following chapter on tree wedding and Irish forestry.
16. Gifford, pp. 74, 177, 305, 404, 433, 465, 482, 587, 593.
17. Hyman, p. 339, qtd. by Gifford, p. 74.
18. Hyman, p. 338, qtd. by Gifford, p. 74.
19. H. Sacher, "A Century of Jewish History," *Zionism and the Jewish Future*, ed. by H. Sacher (London: John Murray, 1917), p. 237.
20. From Charles Parish's "Agenbite of Agendath Netaim" published in 1969, Francis Bulhof's "Agendath Again" (published in *JJQ* in 1970), M. David Bell's 1975 "The Search for Agendath Netaim: Some Progress, but No Solution," R.J.

Schork's "Kennst Du das Haus Citrons, Bloom?," Robert Byrnes's 1992 essay "Agendath Netaim Discovered: Why Bloom Isn't a Zionist," Neil R. Davison's "'Still an Idea behind It': Trieste, Jewishness, and Zionism in 'Ulysses'" and "Why Bloom Is Not 'Frum', or Jewishness and Postcolonialism in 'Ulysses,'" and Wolfgang Wicht's "'Bleibtreustrasse 34, Berlin, W. 15.' (U 4.199)", to name only a few.

21. Relevant publications include Neil Davison's *James Joyce, Ulysses, and the Construction of Jewish Identity: Culture, Biography, and the Jews in Moderist Europe* (Cambridge: CUP, 1996), Merilyn Reizbaum's *James Joyc's Judaic Other* (Stanford: Stanford UP, 1999), and Vincent Cheng's *Joyce, Race, and Empire* (Cambridge: CUP, 1995), among many others.
22. See Richard Ellmann's appendix for *The Consciousness of Joyce,* which includes a bibliographical list of Joyce's library in 1920. Refer in particular to p. 126 of the book.
23. S. Tolkowsky, "The Jews and the Economic Development of Palestine (With Map)," *Zionism and the Jewish Future*, ed. by H. Sacher (London: John Murray, 1917), p. 142.
24. Neil R. Davison, "'Still an Idea Behind It': Trieste, Jewishness, and Zionism in '*Ulysses*'" *JJQ* 38.3/4 (Spring-Summer 2001), p. 387.
25. Dr. S. Brodetsky, "Cultural Work in Palestine," *Zionism and the Jewish Future*, ed. by H. Sacher (London: John Murray, 1917), p. 179.
26. Ibid.
27. Ibid., p. 181.
28. Ibid.
29. Dr. Chaim Weizmann, "Introduction," *Zionism and the Jewish Future*, ed. by H. Sacher (London: John Murray, 1917), p. 8, Italics mine.
30. Brian G. Caraher, "A 'ruin of all space, shattered glass and toppling masonry': Joyce's orientalism in the context of 11 September 2001 and 1922," in *Textual Practice* 18(4), 2004, p. 512.
31. Ibid., p. 512–3.
32. Ibid., p. 513.
33. Tolkowsky, p. 160.
34. Ibid., p. 162–3.
35. Ibid., p. 140.
36. Ibid., p. 143–4.
37. See Abby Bender's *Israelites in Erin*, in particular chapter four and five, for an in-depth elaboration on this issue.
38. Gifford, p. 51–2.
39. Hawkins, p. 46.
40. Laporte, p. 60.
41. Ibid.
42. Ibid., p. 59–60.

43. John Scanlan, *On Garbage* (Bath: Reaktion Books, 2005), p. 43.
44. For more discussions on Viconian philosophy of history and its influence on Joyce, see *Vico and Joyce*, ed. by Donald Phillip Verene (Buffalo: SUNY, 1987). I will also address it in the next chapter on "Cyclops" and afforestation in Ireland.
45. See Rubenstein, pp. 76–80. Here by "public works" I'm borrowing Rubenstein's definition as "a kind of synthetic summation" of his argument, which is "that works of art and public works — here [in his book] limited to water, gas, and electricity — are imaginatively linked in Irish literature of the period for reasons having to do with the birth of the postcolonial Irish state" (Rubenstein, p. 1–2).
46. Rubenstein, p. 20.
47. Matthew Arnold, "On the Study of Celtic Literature," in *Lectures and Essays in Criticism*, ed. R. H. Super (Ann Arbor: University of Michigan Press, 1962), p. 346.
48. Rubenstein, p. 21.
49. Rubenstein, p. 25.
50. David Lloyd, *Irish Times: Temporalities of Modernity* (Dublin: Field Day, 2008), p. 5.
51. Rubenstein, p. 24.
52. Ibid.
53. Ibid.
54. Lloyd, p. 5.
55. Joe Cleary, "Introduction: Ireland and Modernity," in *The Cambridge Companion to Modern Irish Culture*, ed. Joe Cleary and Claire Connolly (Cambridge: Cambridge University Press, 2005), p. 9.
56. Rubenstein, p. 31.
57. Laporte, p. 58–9.
58. Gifford, p. 137.
59. Laporte, p. 61.
60. Gifford, p. 585.
61. Laporte, p. 56.
62. see Gifford, p. 137.
63. Gibson, *Joyce's Revenge*, pp. 1–2.
64. Ibid., p. 2.
65. Hawkins, p. 2.
66. Cheryl Temple Herr, "Joyce and the Everynight," *Eco-Joyce: The Environmental Imagination of James Joyce* (Cork: Cork UP, 2014), p. 41.
67. Letter to Grant Richards, 15 October 1905, *Selected Joyce Letters*, ed. Richard Ellmann (New York: Viking, 1975), p. 79.
68. Letter to Grant Richards, 23 June 1906, pp. 89–90.

69. See Rubenstein's second chapter ("Aquacity: Plumbing Consciousness in Joyce's Dublin") in *Public Works* (Notre Dame: University of Notre Dame, 2011) for detailed elaborations on Joyce's perspective shift on public works from *Dubliners* to *Ulysses,* pp. 43–92.
70. Clive Hart, "Wandering Rocks," *James Joyce's "Ulysses": Critical Essays,* ed. by Clive Hart and David Hayman (Berkeley: University of California Press, 1974), p. 193.
71. Ibid., p. 199.
72. Rubenstein, p. 69.
73. Joe Cleary, "Introduction: Ireland and Modernity," in *The Cambridge Companion to Modern Irish Culture,* ed. Joe Cleary and Claire Connolly (Cambridge: Cambridge University Press, 2005), p. 9.
74. Rubenstein, p. 69.
75. Ibid., p. 71.
76. Hart, p. 193.
77. Hawkins, p. 1–2, my italics.
78. M. Keith Booker, *Joyce, Bakhtin, and the Literary Tradition: Toward a Comparative Cultural Poetics* (Ann Arbor: University of Michigan Press, 1997), p. 41.
79. See Chapter Two: "'Intercourse Had Been Incomplete': Commodification and the Reification of Social Relations in *Ulysses*" in Booker's *Ulysses, Capitalism, and Colonialism,* for example.
80. Hawkins, p. 48.
81. Ibid.
82. Ibid., p. 49.
83. Ibid., p. 52.
84. Ibid.
85. Ibid.
86. For more discussions of the radical politics of gardening throughout modern western history, see George McKay's *Radical Gardening: Politics, Idealism and Rebellion in the Garden* (London: Frances Lincoln, 2011).
87. Hawkins, p. 39.
88. Gifford, p. 585.
89. Significantly, the material he was reading was a contemporary journal called *Tibits* (*Tibits from All the Most Interesting Books, Periodicals and Newspapers in the World*) which started in 1881 and was considered by historians of journalism to mark the beginning of modern popular journalism (Gifford 80).
90. Kiberd, p. 133.
91. Wells, p. 159.
92. Laporte, p. 31.
93. Rubenstein, p. 47.

PART II NATURE, NATION, ECOSYSTEM

3 TREES
History, Performance, and the Viconian Politics of the Forest[1]

I.

> O Ireland! Our sireland!
> Once fireland! Now mireland!
> No liar land shall buy our land!
> A higher land is Ireland!
>
> — James Joyce, V.A.6 Notebook, f. 3v, *Joyce's Notes and Early Drafts for Ulysses: Selections from the Buffalo Collection*, p. 185

James Joyce certainly had the forests in mind when he drafted these verses in the V.A.6 notebook of "Cyclops," although it is unclear as to why he later disposed of the passage. The transformation from "fireland" to "mireland" in the second line corresponds to the progress from ancient land clearings[2] to contemporary Irish landscape of boglands.[3] The Land Purchase Act of 1903 allowed Irish tenants to "buy [...] land," therefore resulting in a more drastic felling of forests in the following decades.[4] Indeed, the history of Irish forestry up to the early twentieth century can be read as a process of struggle between Crown and rebels, landlord and tenants, environmental preservation and industrial profit, and between Anglo-Irish landowners and, after the passing of the Land Purchase Acts, Irish ones. But the question of afforestation[5] is more often than not a more complicated issue than a dichotomy. It has, as Eoin Neeson comments in his introduction to *A History of Irish Forestry*, "paralleled the political history of the country"[6]. In fact, it is an Irish political history of its own.

Natural historians have pointed out the drastic deprivation of trees in Ireland since the time of Tudor conquest. A. C. Forbes, an active advocate of Irish afforestation in the early twentieth century and later the Director of Forestry at the Department of Agriculture and Technical Instruction for Ireland, writes that "the country during the sixteenth and seventeenth centuries was extremely bare of trees, leaving out of account altogether anything in the nature of woods"[7]. Eileen McCracken also notes in her book *The Irish Woods Since Tudor Times* how "[i]n 1600 about one-eighth of Ireland was forested," and "by 1800 the proportion had been reduced to a fiftieth as a result of the commercial exploitation of the Irish woodlands following on the establishment of English control over the whole country"[8]. Such a reduction of Irish woodlands was the consequence of the extensive clearings during the Elizabethan era, the period in which the word "woodkerne" was first coined due to the English fear of the woods as strongholds for Irish rebels. Elizabeth I, well aware of the threat woodlands pose to English settlers, "expressly ordered the destruction of all woods in the country to deprive the Irish of this shelter"[9]. It was a commonly held view at the time that woodlands provide profit (for the English) to earn and shelter (for the Irish) to rally, and for both reasons the woods had to be felled.

In "Cyclops" as well as in his journalistic writing on Galway, Joyce implicitly signals the essential role of timber in Irish trading history. The profits earned from timber trade and the shipbuilding industry significantly influenced the early stage of Irish international relationships. The Citizen in "Cyclops" proudly claims: "We had our trade with Spain and the French and with the Flemings before those mongrels were pupped, Spanish ale in Galway, the winebark on the winedark waterway" (*U* 12.1296-98). It is not too bombastic a statement, since Joyce himself also writes in "The City of the Tribes" that by Cromwell's time "[a]lmost all the wine imported into the kingdom from Spain, Portugal, the Canary Islands, and Italy used to pass through this port [Galway]. The amount imported annually amounted to one thousand five hundred 'tuns', or, in other words, almost two million litres"[10]. But it was not only the commerce of wine that was involved. McCracken points out that during the seventeenth century, "[t]imber and staves were shipped to Scotland, England, Holland, Spain, France, the Canary Islands, and various southern European ports. By 1615 Ireland was sending thirty cargoes of staves each year to the Mediterranean and in 1625 it was said that France and Spain casked all their wine in Irish wood"[11]. The word "bark(barque)," according to the definitions of *Oxford English Dictionary*, usually refers to small vessels but has another meaning of "the rind or outer sheath of the trunk and branches of trees, formed of tissue paralleled with the wood," and even figuratively as the "outer covering" or "external part" of something. Hence the "winebark," apart from meaning the small vessels for wine shipping, may also suggest the "bark," or staves that form the exterior of wine barrels. This is the Citizen's "winebark on the winedark waterway" (*U* 12.1298), where the wine as well as the barrels for wine constituted most of the goods being transported between Ireland and other European countries on the sea.

After the nostalgic reflection on the busy wine trading route, the Citizen in "Cyclops" declares the future revival of Irish marine activities:

> – And with the help of the holy mother we will again, says the citizen, clapping his thigh. Our harbours that are empty will be full again, Queenstown, Kinsale, Galway, Blacksod Bay, Ventry in the Kingdom of Kerry, Killybegs, the third largest harbour in the wide world with a fleet of masts of the Galway Lynches and the Cavan O'Reillys and the

O'Kennedys of Dublin when the earl of Desmond could make a treaty with the emperor Charles the Fifth himself. And will again, says he, when the first Irish battleship is seen breasting the waves with our own flags to the fore, none of your Henry Tudor's harps, no, the oldest flag afloat, the flag of the province of Desmond and Thomond, three crowns on a blue field, the three sons of Milesius. (*U* 12.1300–10)

Most of the harbors mentioned here were "flourishing harbours in the sixteenth and seventeenth centuries"[12] that had fallen from prosperity during the following two centuries, although Killybegs, far from being the "third largest harbour in the wide world," may in fact be "one of the smallest"[13]. The "Galway Lynches" are, as Joyce himself writes of in "The City of the Tribes," one of the prominent tribes in Galway; "the Cavan O'Reillys" and "the O'Kennedys of Dublin" are both powerful houses of Ireland. And by "the earl of Desmond [making] treaty with the emperor Charles the Fifth" (*U* 12.1305), Joyce alludes to the tenth Earl of Desmond, who was "one of the most powerful and independent of the Norman-Irish lords" in the sixteenth century and "bragged that he could field a household army of ten thousand men and launch his own fleet against England"[14]. Although his alliance with Charles V, emperor of Holy Roman Empire and ruler of the Spanish Empire, failed due to his own death, the earl's action against England and his boast that his own fleet rivaled that of England point to the association between marine power and political dominion of Ireland.

It is no coincidence that the Citizen recalls the rebellious fleet of the earl of Desmond in his list of prosperous Irish harbors after he mentions Ireland's business of stave-shipping and timber trade with other European ports. In fact, the rise of naval and sea power strengthened the interdependent manufactures of shipbuilding and timber industry, and in England as well as in Ireland there was "a clear relationship among sea-power, timber, and economic and political dominance"[15]. Especially during the period in which international maritime rivalry was at its zenith, the English government's major aim was rather to secure fresh woodland resources. Robert Greenhaugh Albion writes in *Forests and Sea Power: The Timber Problems of the Royal Navy 1562-1826* that: "The relationship of ship timber to sea-power gave [timber] an importance far above ordinary articles of commerce .... In all the maritime nations of that period, the preservation of ship timber was the chief aim of forest policy"[16]. It is against this background that the woodlands were regarded as a national resource not only economically but also politically and martially.

Joyce makes another reference to the connection between sea power and the Irish woods in the figure of the "M'Conifer of the Glands" (*U* 12.1279–80), father to Miss Fir Conifer, the bride in the tree wedding catalogue (*U* 12.1266–1295). In Don Gifford's note on the potential sources of this figure, Gifford suggests the possibility of its association with Hugh Roe O'Donnell, also known as "O'Donnell of the Glens." In 1601 O'Donnell, along with another Ulster clan chieftain Hugh O'Neill, led a combined force of Irish and Spanish troops to rebel against the English in Kinsale. The failure of the battle resulted in the defeated O'Donnell fleeing, the old Gaelic system of chieftain-rule ending, and the initiation of the Ulster Plantationn. The Kinsale siege transformed the Gaelic administrative system of Brehon Law, brought in Protestants from Scotland and England for plantation, and resulted in the drastic transformation of the Irish landscape. Neeson informs us that "[w]ith the destruction of the short-lived revival of a dynamic Gaelic order, and the Plantation of Ireland, particularly of Ulster, with privileged alien farmers, there began the systematic devastation of Irish forests that led to the substantially forested Ireland of 1600 becoming, by 1750, a treeless wilderness and a net importer of timber"[17].

It was perhaps as early as the time of the Ulster Plantation that the problem of deforestation started to interweave with that of land. The Plantation in Ulstern and elsewhere in Ireland attracted Protestant settlers from England and Scotland, farmers who, in due course, will see to the dominance of Anglo-Irish landlords and their conflict with the Irish tenants during the times of agrarian agitations. These newcomers shared with the Elizabethans a fearful dislike of Irish woods for the same reason that they provided shelter for the Irish troops; identically their preference for the woods' economic profitability induced the excessive felling of forests. Legislation was prompted to protect the profits of the landlord, to regulate the amount of felling allowed for the tenants, or to prohibit woodland clearings; most of it was to little avail. In subsequent centuries that saw an explosion in the use of timber, England was "not to the forefront in European forestry development either in respect of management or legislation"[18]. And while England procrastinated the development of an afforestation industry, Irish forestry was, not surprisingly, neglected.

In 1895 Sir Horace Plunkett, founder of the Irish Agricultural Organisation Society (IAOS), coordinated a "Recess Committee" whose proceedings recommended the establishment of an Irish Agricultural Department. In 1899 the Department of Agriculture and Technical Instruction for Ireland (DATI) was set up, and silviculture (or forest management) was one of its designated functions. In 1902 the Department founded a forestry school on the newly purchased Avondale estate in Co. Wicklow. In 1907 a Departmental Committee of Enquiry was formed by the DATI to investigate the current condition and prospects of Irish forestry. A report as a result of the investigation was presented to the UK Parliament and simultaneously published in Dublin on 6 April 1908. This is the report "of lord Castletown's" which, in "Cyclops," is mentioned by

John Wyse Nolan in his interrupted statement on Irish reafforestation (*U* 12.1258–61). The Right Hon. Lord Castletown of Upper Ossory was among the eight committee members for this inquiry that aims for the improvement of forestry in Ireland. It reports upon the present state of Irish forestry, examines the aftereffects of the enforcement of Land Purchase Acts, considers financial aid and administrative schemes with regard to forestry, and prompts State forestry and co-operative measures to follow.

The passing of the 1903 Wyndham Land Act enabled the DATI and the County Councils to acquire land for afforestation and the preservation of existing forests; however, the lack of funds hindered such acquisition being put into practice. Landowners, old and new alike, saw that potential profits could be earned from selling the standing timber on their properties. The Report points out that "the prevailing spirit of land transfer operations at present is [...] hostile to trees," since "[the] landlord and the Land Commission find them an obstacle to sale, and there is a sort of premium on their removal"[19]. Due to the rush to own land brought about by the 1903 Land Purchase Act, extensive wood-felling ensued.

On the Irish problem of deforestation, in "Cyclops" John Wyse Nolan exclaims:

– As treeless as Portugal we'll be soon, says John Wyse, or Heligoland with its one tree if something is not done to reafforest the land. Larches, firs, all the trees of the conifer family are going fast. I was reading a report of lord Castletown's .... (*U* 12.1258–61)

John Wyse Nolan's sympathy with the re-afforestation campaign and the felled coniferous trees, a couple of lines later, merits the metamorphosis of him into "the chevalier Jean Wyse de Neaulan" who is going to wed "Miss Fir Conifer of Pine Valley" (*U* 12.1267–9). The gigantic guest list for the wedding thence begins:

Lady Sylvester Elmshade, Mrs Barbara Lovebirth, Mrs Poll Ash, Mrs Holly Hazeleyes, Miss Daphne Bays, Miss Dorothy Canebrake, Mrs Clyde Twelvetrees, Mrs Rowan Greene, Mrs Helen Vinegadding, Miss Virginia Creeper, Miss Gladys Beech, Miss Olive Garth, ... (*U* 12.1269–78)

And the list goes on. It is noteworthy that except for the family of the bride (the M'Conifer of the Glands and the two maids of honor, Miss Larch Conifer and Miss Spruce Conifer), none of the attending guests belongs to the coniferous family. Other guests belong to arboreal species other than conifer. It is indeed, like John Wyse Nolan's statement, the procession to witness "the trees of the conifer family [...] going" (*U* 12.1259–60).

Neeson informs us that "[c]onifers were introduced initially towards the end of the eighteenth century in accordance with the new thinking on afforestation in Europe"[20]. They soon became popular since the oak forests of

Britain started to be considered as commercially useless. Then "at no time did conifers supplant existing native hardwoods, by then all but extinct"[21]. In his 1934 article Forbes states that from the mid-nineteenth century onwards, exotic conifers including Douglas fir, Sitka spruce and Japanese larch "have been planted in increasing numbers and are almost familiar to the ordinary observer as the native oak or ash"[22]. The 1908 Committee Report further suggests that "the Forestry Section should adopt as its policy for the first rotation, that of covering the new forest area, as far as practicable, mainly with the quicker-growing varieties of coniferous timber, though mixed, of course, with other species ..."[23]. In the words of the Society of Irish Foresters, "[a]fter 1840 conifers became *the fashion*"[24].

In the tree wedding catalogue it is "[t]he fashionable international world" that "attended *en masse* this afternoon at the wedding of the chevalier Jean Wyse de Neaulan, grand high chief ranger of the Irish National Foresters, with Miss Fir Conifer of Pine Valley" (*U* 12.1266–8). It is also interesting to note that while the arboreal guest list is exclusively composed of female names and titles, some of the titles, notably the three Conifer sisters, are young "Misses." In fact, the felling and export of immature trees was a problematic phenomenon in Irish forestry at the time. One of the industrial witnesses for the Committee Report of 1908 says: "it is not only the mature timber that is coming into the market, but that is swept away and the immature timber is swept away too. Whole woodlands are going down and they don't leave a tree behind, simply the grass is left"[25]. Another witness, who is the representative of a local wood-working industry that uses a great deal of timber, states that "[t]he farms and estates in his district were being cleared of trees," and that his firm were getting "practically all the timber, immature as well as mature, and from both landlords and tenant purchasers"[26]. In other words, many of the trees were felled before they were fully grown. According to the 1908 report, up to 72 percent of the timber cut was exported, and a very large quantity of the exported timber consists of younger trees. Another issue in the timber industry of Ireland is that the trees were not just cut young: worse still, they were sold rough. The report indicates that "practically all of this timber [left] the country in the round state, merely as trunks and logs, without any preparation being done upon it"[27]. This is detrimental to Irish wood-working industries, since the primitive state of timber would reduce its value on the timber market, whereas those shipped from the continent, after being slightly processed by local timber industries, could be sold at a much higher price. In response to the above problem, the committee proposed a co-operative campaign that involves the first stage of State forestry and later the support of State-assisted forest management on private estates.

The 1908 report also singles out the urgent demand for timber supply in the international market as a result of industrial development. It quotes from the Report of the Departmental Committee on British Forestry that "the world is rapidly approaching a shortage, if not an actual dearth, in its supply of coniferous timber, which constitutes between 80 and 90 percent of total British imports"[28]. In Continental Europe, the industrial expansion of Germany has resulted in "an increase of her import of foreign timber from £7,333,000 in 1895 to £13,176,000 in 1905"[29]. It is perceptible that, with the huge demand for timber, forestry and timber export to Continental Europe must have been an industry of much economic potential for the comparatively underdeveloped Ireland of the time. In "Cyclops" the tree-wedding couple's excursion to Germany, "Mr and Mrs Wyse Conifer Nolan['s]" prospective "quiet honeymoon in the Black Forest" (*U* 12.1294–5), may thus suggestively involve a Continental tour of a more progressive afforestation programme.

These facts may help to offer possible explanations for Joyce's writing of the tree wedding catalogue. Afforestation was a popular issue at the turn of the twentieth century, and mass interest was shown in the large number of relevant articles that appeared in both British and Irish newspapers. In the *Freeman's Journal* alone, between 1900 and 1920, there were at least 140 entries whose titles touched on Irish afforestation; 39 were published between the years 1907-08. In the *Times* of London, the number of related articles from 1900 to 1920 was up to 198; 48 of them were released in 1907 and 1908. But afforestation was more than a popular topic in Ireland; it was a much-debated agenda item regarding contemporary colonial politics. It involves the history of plantation and imperialism, and concerns British policies in Ireland. It reflects cultural politics and tenant-landlord conflicts in Ireland over the centuries. And its early twentieth-century supporters include members of the Gaelic League and Sinn Fein, among them Arthur Griffith, who above all "was persistent in crying out for forestry"[30], and Horace Plunkett, who was pressing for the establishment of an Irish institution for its own silviculture. Evidently, forestry is inseparable from the historical process of the construction of Irish nationalism, and Joyce must have been aware of its essential role.

II.

A few weeks after the release of the Committee Report on Irish Forestry, on 22 May 1908, *The Times* published the minutes of the meeting at the House of Commons from the day before. Under the agenda of afforestation, a "Mr. Nolan"[31] from South Louth made a statement about the detrimental influence of British dominion on Irish forests. He "complained that the forests of Ireland had been deliberately destroyed in the past by the British soldiers and afforded shelter to the people" and argued that the British Parliament "owe it to the people to re-afforest their country" (9a). Despite the amusing coincidence that it is also a "Mr. Nolan" (John Wyse Nolan) who, in "Cyclops," campaigns for afforestation, his opinion strikingly echoes Joyce's own comments on this topic. In "Home Rule Comes of Age," first published in the Italian patriotic paper, *Il Piccolo della Sera*, a year previous to Mr. Nolan's statement to the House of Commons, Joyce writes:

> .... Nor will [the Liberal ministers and the opposition newspapers] recall the fact that the politicians and scientists who investigated the vast central bog of Ireland concluded that the two species that sit beside every Irish fireplace, consumption and insanity, are a refutation of all English claims, and that the moral debt of the English government for not having seen to the reforestation of this disease-ridden swamp for over an entire century amounts to over 500 million francs. (*OCPW* 144)

Twelve years later, Joyce recomposed his blunt political statement and transformed it into the Citizen's speech in Barney Kiernan's pub. After a nationalistic reflection on the past prosperity of Irish manufactures, the Citizen imputes the decline of Irish industries to British rule: "What do the yellowjohns of Anglia owe us for our ruined trade and our ruined hearths?" he exclaims, "And the beds of the Barrow and Shannon they won't deepen with millions of acres of marsh and bog to make us all die of consumption?" (*U* 12.1254–57). What Joyce's point, as well as the Citizen's exclamation assume, is the knowledge of a contemporary engineering construction scheme which aimed to manipulate the natural resources of the waste bogland in western and central Ireland. These engineering projects — "designed to deepen the two rivers [the Shannon and the Barrow] in order to drain the marshfield and to develop peat bogs" — aroused considerable public discussion at the time[32]. Significantly, one of the main private investors in these plans was Lord Balfour (the Right Honourable Arthur James Balfour, first Lord of Treasury, leader of the House of Commons, and Conservative prime minister in 1904), who was notoriously hostile to expressions of Irish Nationalist sentiment, and as the result of his coercion policy was nicknamed "Bloody Balfour"[33]. With its policy "to kill Home Rule with kindness," Balfour's government "effectively maintained Liberal policies in Ireland"[34]. It was against this background that "Mr Nolan"'s statement in Parliament, Joyce's political critique in the Italian press, and Arthur Griffith's advocacy for afforestation concurrently emerged.

It is very likely that Joyce draws inspiration from Griffith's Sinn Féin policy for his writing about afforestation, if his interest in the issue didn't stem from the same source. Griffith has been subtly present in *Ulysses*, and critics have signaled

Joyce's persistent attention to him, particularly his early preference for Griffith's non-violent programme and the Sinn Féin movement[35]. Richard Ellmann states that "although [Joyce] refused to endorse the revival of the Irish language, [he] was in other ways on the side of the separatist movement, and particularly of Griffith's programme"[36]. According to Dominic Manganiello, "of all the newspapers competing for the nation's allegiance, [Joyce] singled out the *United Irishman* for particular attention"[37], claiming that it was "the only newspaper of any pretensions in Ireland"(*Letters II* 158) and that "its policy alone would prove beneficial"[38] to the country. Attentive to Griffith's political activities, Joyce was assumingly informed of the former's scheme on afforestation suggested in "The Sinn Féin Policy," which first appeared in the *United Irishman* between January and July 1904, and was later published in the appendix of a pamphlet entitled *The Resurrection of Hungary*.

Joyce didn't keep a copy of Griffith's *The Resurrection of Hungary* in his library as far as we know, but the Irish-Hungarian parallel of the "squareheaded fellow"(*U* 8.463), persuasively argued by Manganiello[39] and Gibson[40], makes Griffith's pamphlet a key text in reading the politics of the nationalist thinking underlying "Cyclops." Andrew Gibson insightfully points out the political context of the Irish-Hungarian analogy in Nationalist circles in Ireland, and how the pamphlet itself, mythologizing the dual monarchy in the Austro-Hungarian Empire, appealed to many Irish nationalists not as history but as "an arousing myth"[41]. Joyce certainly uses much of this republican fascination with the Irish-Hungarian myth in his composition of the historiography and nationalism of "Cyclops," but his treatment of the subject entails much subtler attentiveness to campaign details as well as its rhetorical style.

On November 28th, 1905, at the First Annual Convention of the National Council, Arthur Griffith outlined the Sinn Féin programme. In the policy, he proposed to develop national industry as well as agriculture, advocated a national banking system, and called for an Irish consular service which would help publicise Irish goods. Griffith noted the problem of British taxation and the current law system that devastated the competitiveness of Irish home industries. He claimed that under the Sinn Féin policy "no possibility would be left so far as they were concerned for a syndicate of unscrupulous English capitalists to crush out the home manufacturer and the home trader"[42]. He proposed a prospective scheme of alliance among 159 unions of Ireland to use only Irish flour in the idle mills of the country, and calculated how the plan, when enforced, would advance home industries, decrease Irish unemployment, and bring environmental benefits. Seeing the connection between the national economy and the use of natural resources, he urged the manipulation of Irish wastelands. "Under a National Government," he claimed, "there would be no room for pauperism in Ireland, because under such a Government those unable, as paupers, and those able to work would be provided with it in plenty in reclaiming the four million acres of waste in this country"⁴³. For Griffith, the economic paralysis of Ireland required a national reform in the system of taxation as well as governmental intervention in industrial operations. He saw in the reclamation of Ireland's vast wasteland the hope of reviving the Irish industries of forestry and farming. He sensationally proclaimed that 24 percent of the soil of Ireland "[awaited] the plough or the tree," while meanwhile "the people of the country [were] annually mulcted in millions to keep in soul-destroying prisons those who could carry out the work"⁴⁴. With a question, "Do you ever hear of a free nation paying out its hundreds of thousands of pounds to keep in soul-destroying idleness tens of thousands of its able-bodied population while one-fourth of [the] soil of this country remains awaiting reclamation?"⁴⁵, Griffith forcefully concluded that the only solution to the Irish problem of poverty and unemployment is afforestation. In his opinion, the reclamation of Irish waste lands would contribute to the economic manipulation of the natural as well as the human resources of the country.

James Joyce was aware of the detailed programmes suggested by Griffith's Sinn Féin policy, given the proof shown in his critique of Griffith in his letter to Stanislaus, dated 25 September 1906 (*LLII* 167). In another article he wrote in 1907 for *Il Piccolo della Sera*, he publicised his opinions:

> The new Fenians are joined in a party which is called Sinn Fein (We Ourselves). They aim to make Ireland a bi-lingual Republic, and to this end they have established a direct steamship service between Ireland and France. They practise boycotts against English goods; they refuse to become soldiers or to take the oath of loyalty to the English crown; they are trying to develop industries throughout the entire island; and instead of paying out a million and a quarter annually for the maintenance of eighty representatives in the English Parliament, they want to inaugurate a consular service in the principal ports of the world for the purpose of selling their industrial products without the intervention of England. (*OPCW* 140)

The Sinn Féin stratagems Joyce singles out here: the promotion of Irish linguistic education, the Irish-French commercial connection, boycotts against English products, the encouragement of Irish purchases, and the furthering of the consular system, are all listed in the outline of Sinn Féin Policy. Yet Joyce's interest in the movements advocated in the policy isn't confined to his writing. He, too, was interested in "Foxford tweeds" (*U* 12.1246), which the Citizen includes in his gigantic list of Irish national products just a few lines before the tree wedding passage in "Cyclops." In a letter to Stanislaus, Joyce sympathetically wrote that although "a great deal of [Griffith's] programme perhaps is absurd but at least it tries to inaugurate some commercial life for Ireland" (*Letters II* 167). To show his support of Griffith's policies Joyce too "had aided Irish commerce when he 'took some steps' to secure an agency for Foxford tweeds in Trieste," and when *Sinn Fein* started advertising Irish gowns, Joyce even "offered to buy the patterns if 'Nora liked the style' "[46].

Joyce's awareness of Griffith's Sinn Féin programmes does not, however, stay on the practical level. His attentiveness to the policy is also reflected in his assumption of the compositional style of Griffith's political programmes. After a few lines of contextual analysis of Irish economic paralysis and attention to its wastelands, Griffith emphatically concluded:

> .... The central plain of Ireland awaits only
> **Afforestation**
> to raise the mean temperature of Ireland four degrees, and thus render the soil of Ireland doubly fruitful; and our people are taxed not to carry on so noble a work, but to perpetuate pauperism.[47]

Notice the layout of Griffith's writing: in order to emphasize afforestation as one of his principal policies, Griffith stridently emboldened the word and had it occupy the entire line in the fashion of newspaper headlines. Such a technique would successfully catch readers' eyes, leaving an impression on them with the word "afforestation." It is striking that, Joyce, when composing his tree wedding list in "Cyclops," writes in a stylistic format not unlike Griffith's emphatic composition. Similar to Griffith's way of separating his emboldened keyword (afforestation, in this case) from the text, Joyce's tree wedding catalogue is interposed stridently between the on-going conversations among the barflies, though instead of specifying and emboldening the word itself Joyce makes his invocation to afforestation gigantic.

III.

Whereas Joyce's attitude toward Griffith's Hungarian-Irish parallel remains ambivalent, his fascination with the analogy, as Andrew Gibson registers, derived from a larger political context in which the idea was associated with "Irish republican thinking since at least 1848"[48]. Its influence is reciprocal instead of one-way: Sinn Féin was constructed upon the basis of its Hungarian-Irish parallel, which in turn underpinned Sinn Féin's policies of passive resistance and self-reliance. And "Cyclops," as Gibson persuasively exemplifies, "both plays with and comically repudiates Griffith's identification of [these] two nations, his insistence on their symmetrical relation"[49].

Andrew Gibson regards the tree list as "an outrageously eccentric burlesque of a 'tree-run'"[50]. He traces Joyce's schema of "gigantism" to Standish O'Grady's historical imagination with "gigantic conceptions of heroism and strength" in *The History of Ireland*[51], and contends that there are parodic correspondences between many of Joyce's Cyclopean lists and the Celtic "runs"[52]. Nolan's wedding list, according to Gibson, is among the many comic runs with which Joyce ridicules the style of Revival historiography in his composition of "Cyclops"[53]. However, it is worth taking into consideration that Nolan's wedding is in fact more than a reference to Ovidian metamorphoses or a burlesque of Celtic tree-runs. Its subject is political not only in its references to Irish woodland politics and Griffith's Sinn Féin programmes, but also, though often neglected, in Joyce's characterisation of John Wyse Nolan as "grand high chief ranger of the Irish National Foresters" (*U* 12.1267–8), whose ceremonial parades were often gigantic in scale.

The Irish National Foresters (INF) was established by a few members including Joseph Hutchinson[54] in 1877, in Dublin, under the Friendly Societies Act. It first started as a breakaway from the Ancient Order of Foresters (AOF) as a result of their political incongruities, and afterwards expanded rapidly nationwide between 1899 and 1922. As a benefit society that aimed to promote mutual aid in a community of brotherhood, the society has nothing to do with forestry apart from its title. It retained the AOF's forest analogy to "assist their fellow men who [fall] into need 'as they [walk] through the forests of life'"[55] as its creed, and took "Unity, Nationality, and Benevolence" as its motto. In practice, the Society's fundamental aim was to offer benefits for the burial and sickness of its members; other benefits included elder pensions, unemployment relief and allowances for widows and orphans[56]. Its social concerns interestingly correspond with those of Leopold Bloom's, since the latter enters Barney Kiernan's pub with the single purpose of meeting up with Martin Cunningham in order to negotiate post-funeral arrangements for the diseased Dignam's widow and his children.[57] Previously in "Wandering Rocks," Martin Cunningham and Jack Power appear outside the City Hall building fundraising for the Dignams, and John Wyse Nolan, our INF man, stays with them.

Although the INF Society declared itself to be non-political and non-sectarian, its congregations, parades, and banners spoke otherwise. Anthony D. Buckley comments that some of these friendly societies, INF included, operated under the guise of the National Insurance Act (1911) to further their own political aims[58]. In fact, the initial 1877 turmoil that resulted in the separation of the INF from the AOF entailed some of its Irish members' attempted commemoration of the Manchester Martyrs. In the 1890s police sources "expressed alarm at the extent to which the INF had been infiltrated by extremists," and in 1890 "a significant section of the organization was dominated by advanced nationalists such as the IRB"[59]. Suspecting its righteousness as a benevolent society, Captain H.B.C. Pollard, a British Staff Officer of Intelligence who served in Dublin during the War of Independence, even claims there is no doubt "that some [INF] members who hold office in the Society use the Society as a cover for criminal societies with which they are personally connected"[60].

By 1911 the Irish National Foresters was the largest friendly society in Ireland, and its nationalist sentiment can be detected simply by its branch names in Dublin and the north of Ireland as of 1919: Sarsfield, Parnell, Brian Boru, Napper Tandy, Thomas Russell, Robert Emmett, William Orr, John Mitchel, along with some Catholic saints and religious leaders among its popular branch titles[61]. Its nationalistic sentiments can be deduced by its branch names and public demonstrations, not to mention its inevitable affiliation with Catholic institutions and activities as a result of its members' exclusively Irish descent. The usual route of the Forester parades would take its participants "to a local church for Sunday mass in the midst of the weekend convention"[62], with a priest's speech, either in the middle, or at the end, of the convention. Such parading of course reminds the reader of the tree run in "Cyclops" which, coincidentally, also ends with a church service (*U* 12.1288–94).

The INF society was engaged in Irish nationalist politics not only in its branch names but more practically in nationalist programmes. Campbell indicates that the INF branch in Newry played an essential part in assisting the establishment of the Gaelic League in the town; in April 1907 alone "the sum of £1 was donated and the branch pledged its support to the League in its work in the preservation of the National language"[63]. In a *Freeman's Journal* issue in 1909, voices from INF branches emerged to "demand that the teaching of our national language be made compulsory in the new University" with the belief that the "grand old tongue should get its rightful place on the programme, and that the University should be a real Irish and National one". By the same year, it even became compulsory that all ranked INF members have knowledge of the Irish language[64]. It is pertinent to recollect John Wyse Nolan's concern with the Irish language in *Ulysses*. In "Wandering Rocks," John Wyse Nolan remains curiously silent on the subject. In "Cyclops," as soon as he enters Barney Kiernan's pub, the citizen enquires of him regarding the latest information on the meeting held at the city hall to "decide about the Irish language" (*U* 12.1181–2). In return, John Wyse Nolan replies in a language that is satirically rendered into a silent movement in mock-heroic word-play:

> O'Nolan, clad in shining armour, low bending made obeisance to the puissant and high and mighty chief of all Erin and did him to wit of that which had befallen, how that the grave elders of the most obedient city, second of the realm, had met them in the tholsel, and there, after due prayers to the gods who dwell in ether supernal, had taken solemn counsel among mortal men the winged speech of the seadivided Gael. (*U* 12.1183–89)

In other words, he raised the issue in front of the Dublin officials and solemnly counselled with them during the City Council Meeting. Joyce uses mock-heroic language to transform Nolan's speech into a pseudo-heroic satire, and later taunts Nolan's advocacy of the Irish language again, when he makes Nolan's name into a Frenchified version — Jean Wyse de Neaulan (*U* 12.1267). Fritz Senn persuasively associates the Frenchification of John Wyse Nolan's name with the scene of Haines's encounter with the Irish milkwoman in "Telemachus" (*U* 1.422–29), where "Irish is actually spoken" rather than "merely clamoured for in patriotic display," yet is nonetheless mistaken for French[65]. The name of the tree-marrying bridegroom being Gallicised into Jean Wyse de Neaulan, therefore, may have more to do with his enthusiasm for the Irish language than with any Continental connection.

In response to Nolan's report on the City Council meeting, the citizen says: "It's on the march. .... To hell with the bloody brutal Sassenachs and their *patois*" (*U* 12.1190–91). Regardless of the connotational meaning of "the march" as the progress of the Irish language debate, there are in fact a few literal "marches" throughout the episode — not least among them is the tree marching led by John Wyse Nolan himself, defender of the Irish language, high chief ranger of the Irish National Foresters. Indeed, apart from the INF's designated function of brotherly welfare and its enthusiasm for the language movement, it was in the occasional street parades that the Society's communal sense of nationalism was publicly displayed (see Fig 3.1). The Foresters gave such weight to their marching demonstrations that they, in Neil Jarman's phrase, "[upheld] the tradition of friendly societies as parading bodies"[66]. Jarman informs us that the contemporary Foresters' parades in Northern Ireland are held mainly on commemorative nationalist occasions including Bloody Sunday, St Patrick's Day, Easter Rising, on days in memory of the Hunger Strikes and Wolfe Tone, and for the INF Annual Parade that takes place each year on the 1st of August[67]. Presumably, the marches INF members participated in the early twentieth century would entail at least St Patrick's Day and its mid-year Annual Parade. Irish newspapers reported these parades in minute detail, describing its "very striking and impressive appearance" with "scenic effect". The 1906 *Irish Independent* recorded the INF St. Patrick's Day demonstration as a "STATELY PARADE" with "some 1,000 numbers". In 1923, the INF Annual Convention parade in Trinity College Dublin involved such a

Fig. 3.1: Crowds of Spectators Observing a Parade Passing Through O'Connell Street. Picture taken by Clarke, J. J. Published between 1897–1904. Courtesy of the National Library of Ireland. <http://catalogue.nli.ie/Record/vtls000168779> Accessed 16 June 2012.

large number of its participants that "Parnell Square could not hold the crowd taking part"[68].

Typical early-twentieth-century INF parades unexceptionally comprise leading brass bands (such as the one in Fig 3.2), a significant number of members in green regalia, large illustrated banners of its branches, and female representatives in green costumes. A descriptive account in a 1915 issue of *Freeman's Journal* exemplifies such a procession on St. Patrick's Day:

> .... More than five hundred members of the Order took part in the procession, accompanied by their brass and reed band, and the St. Kevin band, and some beautiful banners. A large number of members wore the picturesque costume of the Order, and several lady members, wearing green cloaks, formed a conspicuous feature of the parade, which was well marshaled, and was watched by large crowds as it passed through the streets.

Fig. 3.2: INF brass band in Monaghan, 1910. Picture courtesy of Trinity College Library Dublin.

This picturesque green procession "on the march" (*U* 12.1190), I argue, would be the "tree run" Joyce writes of as one of the multiple layers of the gigantic tree catalogue in "Cyclops," and the leading brass band of the parade (see Fig 3.5.), Joyce's mocking rendering of the "Green Hungarian Band." D. P. Moran, Griffith's counterpart in Irish nationalist movements, does not buy into Griffith's symmetrical Hungarian-Irish analogy, and instead reputedly mocked the parallel by nicknaming Sinn Féin "the Green Hungarian Band"[69]. Infamously hostile toward the Unionist and non-Catholic revivalists, Moran has been associated with Joyce's portrait of the Citizen due to his notorious language. Emer Nolan reminds us that: "[Moran]'s own satiric portrait of the corrupt Irish nationalist in *The Philosophy of Irish Ireland* closely resembles Joyce's representation of the hypocritical rhetoric of the Cyclops"[70]. This

"Green Hungarian" nickname for Sinn Féin is not only playfully correspondent to a later scene in "Cyclops," where Bloom as a Hungarian hero is sent off with a band playing the unofficial Hungarian state anthem, *Rakóczsy's March* (*U* 12.1814–42). In fact, unnoticed by Joycean critics thus far, the tree wedding catalogue of "Cyclops", shadowed by Griffith's politics and politically performative in green demonstrations, itself appears very much like Joyce's mocking version of the "Green Hungarian Band."

The Foresters' parade was influential not only in newspaper reports. It was, for citizens of Dublin, a spectacular scene of the urban streetscape. C. P. Curran, Joyce's colleague at the University College Dublin, has a vivid account of the Foresters' "picturesque" parade in his book *Under the Receding Wave*. These Irish National Foresters, in Curran's words, were no less impressive even compared with the "resounding and brilliant spectacle" of Dublin's Fire Brigade, whose marching grandeur outshone the Lord Mayor's official convoy[71]. He reflects in detail the Foresters' procession in which

> [t]he scarved members were preceded and flanked by a splendid cavalcade in the Robert Emmet costume which coloured lithographs had made no less familiar than St Patrick's. They wore dark green, wide brimmed felt hats with white plumes, green cutaway coats with epaulettes and gilt buttons, white knee-breeches, and riding boots. At their head, mounted on a Shetland pony, rode a little girl with golden ringlets, dressed in the same attire. Year after year the Dublin crowd waited for her apparition, changed but unchanged. They cheered her, loudly recognizing any singularity in her followers; loved, laughed at and admired their gallant incongruity.[72]

Such was the picturesque appeal of the Foresters' parades. In the tree catalogue in "Cyclops," John Wyse Nolan (in his mock-Gaelicised name), our "grand high ranger of the Irish National Foresters," emerges with a gigantic arboreal crowd whose presence is presumably green and "foresty." Nolan's bride dresses in "green mercerised silk," "sashed with a yoke of broad emerald" and "bretelles and hip insertions of acorn bronze" (*U* 12.1280–84), not dissimilar to the INF members in their green regalia and their "green cutaway coats with epaulettes and gilt buttons"[73]. Like C.P. Curran, Joyce must also have sensed the "gallant incongruity"[74] of the Irish National Forester parade. The banners these Forester marchers carried constitute another facet of the Society's picturesque nationalist demonstration. In sensual effect, these banners contribute "a sense of a floating existence" to the entire procession[75]. They were gigantic in size, "usually supplied complete with 12ft stained and polished carrying poles, fitted with brass ferrules and ornamented spearheads"[76]. These banners were painted with figures of Irish nationalist leaders, patriotic rebels, and more dominantly, images of Erin (see Figs. 3.3, 3.4, and 3.5). Campbell describes these substantial figures of Erin

Fig. 3.3: INF banners with figures of Erin. Courtesy of Trinity College Library Dublin.

assertive with unfurled flags, wearing "a pure white dress, ... enclosed by a green cloak"[77]. These Irish outfits may have given inspiration for the female marchers' costumes at the Forester parade, and possibly, for Miss Fir Conifer's wedding dress (*U* 12.1280–84).

Fig. 3.4: INF banners with figures of Erin. Courtesy of Trinity College Library Dublin.

Perhaps not too surprisingly, dark green was not the color exclusively of the Irish National Foresters in early-twentieth-century Ireland. Cumann na mBan, whose activities mark women's participation in republican movements from 1914 onwards, has the real-life John Wyse Nolan's wife, Jennie Wyse Power, as one of

Fig. 3.5: INF members holding a banner of theirs, Ballyholland, August 1995. Seen in Jarman, *Material Conflicts,* p. 30.

its founder members as well as its first president. These female Republicans of Cumann na mBan, too, dress in dark green.

In fact, the real-life John Wyse Nolan, John Wyse Power, is not known to be associated with the Irish National Foresters, but was best known for founding the GAA (Gaelic Athlete Association) with Michael Cusack, and was reputedly a member of the IRB (Irish Republican Brotherhood).[78] With his wife Jennie, John Wyse Power co-established "The Irish Farm and Produce Company," which is to become, in *Ulysses*, "that Irish farm dairy John Wyse Nolan's wife has in Henry Street" (*U* 8.950–51). Located on 21 Henry Street, Dublin, the shop and restaurant served as a popular meeting place for many of the cultural and political organizations with which the Wyse Powers, Jennie in particular, were involved. An influential journalist, John Wyse Power worked as an editor for the *Freeman's Journal*, the *Irish Daily Independent*, the *Evening Herald*, and the *Evening Telegraph*. He was an active Gaelic Leaguer, served as a member of the executive of Cumann na nGaedheal for many years, and was "an outspoken opponent of the royal visits of the early to mid-1900s, allegedly being the author of the popular Dublin expression, 'kiss my royal arse'"[79]. Whereas the expression is nonetheless taken by Myles Crawford, Joyce's fictional editor of *Evening Telegraph*, when he rudely and impatiently replies to Bloom's request for the Keyes ("He can kiss my royal Irish arse, Myles Crawford cried loudly over his shoulder.") (*U* 7.991–2) in "Aeolus," John Wyse Nolan shares with John Wyse Power the nationalist disdain for royal marches with his "cool unfriendly eyes" (*U* 10.1036) at the sight of the viceregal cavalcade. Despite the absence of the INF connection, the political attitude of John Wyse Power corresponds to that of John Wyse Nolan in their common support of the Gaelic League and their contempt for royal parades. Furthermore, both John and Jennie Wyse Power were well acquainted with Arthur Griffith. More fervent in political activities than her husband, Jennie even represented the national Council of Arthur Griffith as a member of the board of guardians of North Dublin Poor Law Union from 1903 to 1912[80].

Considering the Wyse Powers' enthusiastic pro-Griffith involvement in nationalist activities, the tree wedding scene in "Cyclops" can be read not only as a mock-Celtic tree-run but moreover as a "tree-run" referencing contemporary republican movements. In the preface to *Beside the Fire*, Douglas Hyde explains the Celtic "runs" with an example from "The Slender Grey Kerne," in which "the swift movement of the kerne" is described[81]. In other words, the style of "tree-run," as Hyde coins it, is adopted to depict the kerne's rapid movements; relative to the movement of the kerne, the trees seem to run away from him. The "tree-run," therefore, is not so much a personification of the trees as the relatively rapid motion of the kerne. Bearing in mind my previous elaboration on the specific political and colonial context of the phrase "woodkerne," the gigantic "tree-run" in "Cyclops" can possibly be regarded more specifically as a "woodkerne-run," in which not only the nationalist Irish National Foresters, but more significantly the female revolutionists of Cumann na mBan, in their green uniforms, march.

Historical and anthropological studies of Irish national symbols have signaled the prominence of a passive female figure, namely Erin, on nationalist prints, products, advertisements and, pertinent to our purpose, on banners of nationalist organizations such as the INF[82]. I have previously illustrated that the tree wedding catalogue entails the performative function of the nationalist movements by mockingly arborizing the INF paraders and female nationalists of Cumann na mBan, while it also puts the debate about Irish re-afforestation on the show. However, the almost exclusively female names of the tree wedding list in "Cyclops" suggest implicit violence in its performativity. This gender-specific catalogue arouses the question of patriarchy in the discussion of Irish nationalism, and, I argue, discloses an alternative historiography of Irish patriarchal feudalism. Such violence of feudal patriarchy in the socio-historical context of Irish nationalism, as I will elaborate further in the following section, owes to Vico's *New Science* not only the latter's natural law philosophy but even more exquisitely the tree catalogue's very location within the episode of "Cyclops" as the rattling beam in the Cyclopean eye.

IV.

> *This was the order of human institutions: first the forests, after that the huts, then the villages, next the cities, and finally the academies.*
>
> — Giambattista Vico, *New Science* § 239

Brigitte L. Sandquist's 1996 article that appeared in *James Joyce Quarterly* may be among the few scholarly essays that solely and extensively examine the Tree Wedding catalogue in "Cyclops," although numerous critical comments have been made with regard to its gigantic style as exemplary of the episode. Among others, Fritz Senn in "Ovidian Roots of Gigantism in Joyce's *Ulysses*" singles out its possible sources from Ovid's *Metamorphoses*[83], from where Sandquist develops her analysis on the tree-human transformation and the implicit violence of patriarchal space in the first part of her article. She then turns toward the structural dynamics of the catalogue within the narrative as a whole and contemplates its thematic eye/I distinction. She parallels the Homeric design of the eye-blinding scene with Joyce's tree wedding catalogue that is situated right between two "eyes" (*U* 12.1265, 1296), and interestingly juxtaposes Odysseus's alienness in the context of the Cyclops' land with the tree wedding's alienness in the progress of the Cyclopean narrative. Sandquist sees the interpolation of the catalogue as an alien intrusion into the narrative, and signals the performative function of the tree catalogue as it literally "hits the narrative right between the eyes"[84].

Situated between "Europe has its eyes on you" (*U* 12.1265) and "And our eyes are on Europe" (*U* 12.1296), the catalogue not only textually becomes the Cyclopean eye in Joyce's contemporary Irish context, but more significantly, it is one with the "beam in the eyes." Joyce's use of the beam-in-the-eye metaphor is however far beyond the synchronic structure of the catalogue. In fact, the episode registers a philological historiography which, as critics have observed, Joyce "absorbs"[85] from the work of the eighteenth-century philosopher Giambattista Vico. In what follows, I will illustrate how he further internalizes the Viconian philological historiography and reconsiders the systems of law, patriarchy, and the institutionalization of the nation, specifically in an Irish context, and even more specifically in the tree wedding catalogue.

The name Giambattista Vico is not a foreign one to most Joycean readers. Much attention has been directed toward *Finnegans Wake*, which Joyce himself advises Miss Shaw Weaver to read along with Vico's *New Science*, the way he recommends *The Odyssey* to be read as a counterpart to *Ulysses* (*JJ* 564). Some Viconian traces have been observed in *Ulysses,* and it is also noted that "[the novel's] Odyssean parallelism may have been partly influenced by Vico's 'discovery of the true Homer'"[86], which bears the same title as the third book in the *New Science*. A few scholars have commented on Vico's influence on Joyce not only in his composition of *Finnegans Wake* but also of *Ulysses*, and it is frequently noted that Joyce confesses that his "imagination grows" when he reads Vico as "it doesn't when [he] read[s] Freud or Jung" (*JJ* 693). A. Walton Litz lists "Cyclops," along with "Nestor," as the two *Ulyssean* episodes with the most explicit Viconian influence[87]. As early as the 1930s, Stuart Gilbert indicated *Ulysses*'s Cyclopean allusion to Vico's giants. He refers Joyce's technique of "gigantism" in the twelfth episode to the Viconian mythological historiography, noting that in Joyce's "Cyclops," "not only is there an elaborate description of a giant but the technique of the episode itself is, as we have seen, 'gigantism'"[88]. Gigantism, Joyce's designated episodic technique for "Cyclops," can hence be regarded not simply as a Homeric or an Ovidian scheme, but also, if no less significantly, a Viconian one.

Vico's giants appear in Book Two of the *New Science*, where he explains the development of human civilization from savagery, the constitution of the family, to that of the nation. In his view, after the flood, Noah's descendants, due to their primitive lifestyle in scattered forests over the earth, have grown into giants. They live unaware of divine Providence until the thunder strikes, and, out of corresponding awe and fear for the divine, they stop copulating, hide in the caves and start raising families. This, according to Vico's theory, marks the beginning of religion (fear of god) and matrimony (the institution of monogamy under divine surveillance). From then onwards, families develop their genealogical lineage, settle in cities, and generate the social system of laws and patriarchy.

Although Joyce's Viconian adoption is most notably demonstrated in the incarnation of circular progression and cyclical history in *Finnegans Wake*, Vico's *New Science* is in truth a political text, since, as Samuel Beckett points out, "more than three-fifths" of the book is "concerned with empirical investigations"[89]. Stuart Gilbert, too, specifies that "the theory of recurrence in the affairs of men and nations" in *Ulysses* is "not peculiar to mystical thinkers [of the Celtic twilight]; it appears as an empirical deduction from the facts of history, rather than as an *a priori* dogma, in the works of the Italian philosopher Vico"[90]. The book's full title itself, *Principles of New Science of Giambattista Vico concerning the Common Nature of the Nations*, endorses its political aim. It is structurally divided into five sections based on the Viconian plan of cyclically developmental history of humanity. Apart from Book One, which introduces the establishment of principles, and the last one that marks the historical cycle, the entire piece runs according to Vico's three successive stages of progression: that of gods, of heroes, and of men. His discourse of historiography attempts to incorporate diverse myths and legends by narrating a new universal history that dwells on the origin and progression of poetry, language, law, institutions, and human civilization. And it is this "theory of the origins of poetry and language, the significance of myth, and the nature of barbaric civilization," as Beckett claims, that "appears the unqualified originality of [Vico's] mind"[91]. Joseph Mali even resolutely asserts: "It was Vico's analysis of the inner historicity of human institutions — as exemplified, for instance, in the etymological explanation of words or in the psychological conception of theogony as the outcome of traumatic experience — which […] impressed Joyce so profoundly"[92].

In a striking passage demystifying the myth of the one-eyed Cyclops, Vico writes:

> Every clearing was called a *lucus*, in the sense of an eye, as even today we call eyes the openings through which light enters houses. The true heroic phrase that "every giant had his *lucus*" [clearing or eye] was altered and corrupted when its meaning was lost, and had already been falsified when it reached Homer, for it was then taken to mean that every giant had one eye in the middle of his forehead. …. (*NS* 564)

Here Vico associates *lucus,* clearing, with the concept of an eye. His explanation for the Cyclopean myth corresponds to his earlier axiomatic "universal principle of etymology in all languages" that "words are carried over from bodies and from the properties of bodies to signify the institutions of the mind and spirit" (*NS* 237). Later he brings up the idea that "in all languages the greater part of the expressions relating to inanimate things are formed by metaphors from the human body and its parts and from the human senses and passions" (*NS* 405). Such physical personification of language and geography, Rosa Maria Bosinelli maintains, is the Viconian principle that "inspires the very first page of *Finnegans Wake*; a page that declares its debt to Vico from the opening word 'riverrun'"[93]. It is very possible that Vico's account of the triangular *lucus*-clearing-eye analogy also inspires Joyce's rendering of the "tree-run" in "Cyclops." Sandquist's analysis has already reminded us of the performativity of the catalogue with the tree list situated between the two "eyes" within the entire episodic framework. Bracketing the tree catalogue, the narrative would appear as follows:

– Europe has its eyes on you, says Lenehan.
[tree wedding catalogue]
– And our eyes are on Europe, says the citizen. (*U* 12.1265, 1296)

Sandquist concludes with the notion that the above structure is rhetorically equivalent to Bloom's previous statement: "some people ... can see the mote in others' eyes but they can't see the beam in their own" (*U* 12.1237–8). She also notices "implicit violence in this catalogue, a violence masked by Joyce's parodic use of the society column," and how a rhetorical violence has been committed to shatter "both the *vision* and the *subjectivity* of the narrative"[94]. However, it is worth noting that the specific rhetorical structure of Joyce's tree catalogue may actually be visually and subjectively coherent with Vico's philological interpretation of the Cyclopean myth.

A closer investigation into the political context where Vico sets up his *lucus*-clearing-eye analogy helps to explain the violent undertone of the Cyclopean myth in both Viconian and Joycean settings. After the development of families and the agricultural institution of fathers, Vico introduces the emergence of "*famuli*" who, originally as savages beleaguered by others, came under the fathers to seek asylum. This new social group as a feudal underclass subsequently grew in number to accommodate martial slaves and refugees, and, many of its social rights denied, was subjected to unequal servitude. They lived with restricted bonitary rights under the "eternal principles of the fiefs" (*NS* 599), and were denied land possession. With the anger of the repressed, in due time these *famuli* rebelled against the fathers. It is against this feudal background that the confrontation between the Roman plebeians (the *famuli*) and the patricians (the fathers) came to light, the confrontation which, as Thomas Hofheinz observes, is not unlike the history of Irish-British conflicts which Joyce understood[95]. In *Joyce and the Invention of Irish History*, Hofheinz calls for critical attention to Vico's primary avowed purpose in his composition of the *New Science*, which is "to construct a provincial historiography that would lay the true groundwork for the teleological explanation of political power advanced by various proponents of 'natural law' philosophy"[96]. He persuasively argues that "Joyce recognised in Vico's providential history an anatomy of modern authoritarian forces dominant since the rise of feudalism," and that Joyce's "parodic treatment of that history in *Finnegans Wake* reflects his sensitivity to [the] ways in which Vico's patriarchal natural law, within the providential scheme, justified England's conquest of Ireland and clarified the complex power politics beleaguering modern Irish experience"[97]. Since Joyce became Vico's "enthusiastic" reader as early as in his Trieste days in 1913-1914 (*JJ* 340), he must have already perceived the analogous relation between Irish historical experience and Viconian historiography before he started composing *Finnegans Wake*. The de-mythologized Cyclopean eye, I propose, is among the Viconian "poetic imaginations" which Joyce "absorbs" to construct his own re-mythologized historiography of Irish feudalism and nationhood in "Cyclops."

After the appearance of the *famuli*, which "preceded the cities, and without which the cities could not have been" (*NS* 553), Vico proceeds to illustrate its significance in the development of human civilization. As the *famuli* gather under the asylum of the heroic families to "guard and defend each his own prince and to assign to his prince's glory his own deeds of valor" (*NS* 559), there was feudalism initiated, followed by the "first intimidation of the fiefs" (*NS* 557). Vico considers the formation of arms and shields inseparable from the crucial feudal position of the *famuli,* whose alternative name, *clientes,* originally means "to shine in the light of arms (whose splendor was called *cluer*)," because "they reflected the light of the arms borne by their respective heroes [in their families]" (*NS* 556). He associates shields with the clearing of fields, signaling "in the science of heraldry the shield is the ground of the arms" (*NS* 563). These first shields contain colors denoting fields of arms, the colors which, in black, green, gold, blue, and red, are believed to relate to the cultivation of land (*NS* 563). And according to him, these first shields were round in shape because "the cleared and cultivated lands were the first *orbes terrarum* [the circle of the lands]" (*NS* 564). It is in this context that he elucidates the *lucus*-clearing-eye analogy, and how the heroic phrase "every giant had his *lucus*," which originally describes the act of land clearing by the giants, is misunderstood when *lucus* is mistaken to mean "eye" instead of "clearing"—hence the myth of the one-eyed Cyclops, made renowned by Homer.

In *Ulysses*'s "Cyclops," the conversation taking place in Barney Kiernan's pub reaches the debate on afforestation where the gigantic tree wedding catalogue emerges. I have already elaborated on the catalogue's implicated politics of forestry in Irish history, and how the afforestation movement, propagandized by

Irish nationalists specifically exemplified by Arthur Griffith, finds its way into the narrative to signify contemporary environmental politics as well as the political environment of the time. The catalogue's textual location between the "eyes," as Sandquist contends, corresponds to the Homeric "beam-in-the-eye" myth of the Cyclops. But in Vico's demystified philological account of the Cyclopean myth, the term *lucus* is not the single eye on the forehead, but rather the land clearings, that these giants have. In the tree wedding catalogue we have personified "beams" between the eyes, but these beams, echoing the call for afforestation in early twentieth-century Ireland, act as witnesses to the country's deforested landscape as well as its afforestation propaganda. While the history of Irish woodlands, like the universal law narrated by Vico, initiates with ancient land clearings under feudalism, the "clearings" between the eyes in "Cyclops" may respond to more than simply a dichotomy between the environment and civilisation, between periphery and empirical centre, or between Irish nationalism and British capitalism. They entail a more dynamic philological examination of history that originates in etymology, the origin of language, to recover the Viconian historiography which "goes round and round to meet where the terms begin" (*FW* 452.21–22). While Vico "looked upon language as fossilized history and sought to recover the past from the radical meanings of words," Joyce "reversed this process and sought to create new verbal units which would embody the entire history of a theme or motif"[98]. In this way, as Vico recovers the mystification of the one-eyed Cyclopean myth, Joyce certainly makes use of the demystified *lucus* as land clearings by re-mystifying the clearing (*lucus*), connecting it with the one-eye myth of Cyclopes, and locating it between the eyes in his episode of "Cyclops."

Following his demystification of the myth of the one-eyed Cyclops, Vico continues to illustrate these giants' act of clearing, in association with the antagonist Vulcan, in the sense of rebels. He writes: "With these one-eyed giants came Vulcan to work in the first forges — that is, the forests to which Vulcan had set fire and where he had fashioned the first arms, which were the spears with burned tips" (*NS* 564). In Joyce's "Cyclops" the lethal spear enters the narrative in the very beginning of the episode when the sweeper "near drove his gear" into the anonymous narrator's eye (*U* 12.3). Later the beam-into-eye allusion reappears when the tree wedding catalogue, as a gigantic list of personified beams, intrudes into the narrative between two "eyes." These beams, sharpened into spears, become for Vico the earliest weapons and for Homer the instrument with which Odysseus blinds the Cyclops' eye. In the Homeric myth we are familiar with, Odysseus chars the tip of the olive pike before he and his comrades ram it into the giant's eye socket. In Viconian historiography, the charred spear is Vulcan's weapon, equipped with the cleared forests to observe Jove from the cleared ground and to revolt against him. The way Vulcan burns the forest in order to set armies against Jove is the opposite of the Irish woodkerns, who use woodlands to secure their arms against their enemies from Elizabethan colonizers onwards. Yet remember the tree catalogue's political context within Irish nationalist movements which I have previously mentioned: the Irish National Foresters, the non-militant nationalist society in which the fictional bridegroom is actively involved, and Cumann na mBan, the paramilitary organisation composed of female republicans, including John Wyse Nolan's real-life partner Jennie Wyse Nolan, are both active nationalist organisations at the time. Cumann na mBan, needless to say, was active in republican revolutionary movements. And whereas the INF is not known to participate in armed nationalist activities, the society demonstrates its pro-nationalist claims in press pronouncements and public street parades. Some of its members were also involved in the 1916 Easter Rising, though due to their double membership in the IRB (Irish Republican Brotherhood), their INF identities were less known to the public. The rebellious nationalist undertone in the tree catalogue's "woodkerne-run," therefore, is not unlike the rebellious Vulcan antagonistic to the authoritarian power of Jove, and for the Irish nationalists, the British sovereignty.

V.

Vico lists matrimony, together with religion and burial, as the first three human institutions of civilization. Throughout *Ulysses* we have countless allusions to the Church, not to mention Stephen's much-quoted notion that he has two masters, one of them Italian, by which he means the Roman Catholic Church (*U* 1.638). The "Hades" episode of *Ulysses* is devoted entirely to a funeral, whose influence remains in the citizens' thoughts and conversations throughout the rest of the day. On the other hand, there's another dimension of the burial of the dead that is linked to the Celtic Revival's aspiration to ancient heroism. Such heroic aspiration is exemplified in the mock-bardic catalogue of "Inisfail the fair" near the beginning of "Cyclops," where there are, in "the land of holy Michan," "the mighty dead" who sleep "as in life they slept, warriors and princes of high renown" (*U* 12.69–70). Vico believes that Roman law "calls for burial of the dead in a proper place" (*NS* 531), therefore making the ritual of burial a religious act. Furthermore, from the idea that "by the graves of their buried dead the giants showed their dominion over their lands" (*NS* 531), the burial ground becomes the terrain in which the genealogical tree germinates. I am not suggesting that every reference to a burial or funeral in *Ulysses* must refer to Vico's theory in the *New Science*. However, burials in general, specifically family burials, pave the ground for genealogical lineages. Vico points out that the heads of the Latin families "called themselves *stirpes* and *stipites*, stems or stocks, and their progeny were called *propagines*, slips or shoots" (*NS* 531), namely youthful branches. Hence comes the ancient heroic saying: "we are sons of this earth, we are born from these oaks" (*NS* 531).

Let us return to the tree wedding in "Cyclops," where the act of burial is absent, but the image of an arboreal family indicative of a genealogical tree is not simply present but gigantically dominant. Considering Vico's three institutions of civilization, it is curious to notice the scarce presence of the marriage ceremony in *Ulysses*, regardless of the book's numerous references to the other two Viconian "institutions." In fact, apart from the absence of nuptial rituals, the failure of marriages seems to dominate the book. One of the central storylines of *Ulysses* consists of a strained marriage which leads to the hero's reconciliation with his wife's infidelity. And neither Father Conmee's thoughts on "the hands of a bride and a bridegroom, noble to noble, … impalmed by Don John Conmee" (*U* 10.177-8), nor Bloom and Stephen's mock-nuptial coupling figures "*to be married by Father Maher*" (*U* 16.1887-8), seems to offer a satisfactory version of conjugal relationships. Indeed, the marriages mentioned in *Ulysses* are treated either with a sarcastic undertone or endowed with failure and paralysis. Robert Spoo rightly suggests that Joyce rejects the institutionalization of marriage in his writings as well as in his personal relationship with Nora Barnacle, whom he didn't marry until 1931, after more than two and a half decades of cohabitation. For Joyce, marriage is "complicit with an oppressive political power system"[99], which he solemnly opposed. In a letter to Stanislaus, he commented on the news of Oliver Gogarty's marriage that "Gogarty would jump into the Liffey to save a man's life but he seems to have little hesitation in condemning generations to servitude" (*Letters II* 148). Joyce sees the matrimonial institution as a part of the suffocating patriarchal system that he, when fleeing from Ireland, has attempted to free himself from. For him, it is a demanding tyrannical power that forces married men into "servitude," the servitude which he refuses to fall into, the way Stephen refuses to be the servant of his masters. With regard to Joyce's manifest hostility toward marriage, Spoo concludes: "No one with opinions like these could have written a novel concluding, as *Emma* does, with the assurance that 'the predictions of the small band of true friends who witnessed the ceremony, were fully answered in the perfect happiness of the union'"[100].

By the same token, the wedding ceremony between Jean Wyse de Neaulan and Miss Fir Conifer of Pine Valley cannot possibly be a manifestation of such an Austenian marriage ending in "the perfect happiness."

Joyce doesn't put his tree wedding ceremony into the journalistic language of advertisements in the newspapers simply to "reflect" or "parody" the popular culture. Rather, I argue, in this catalogue he intentionally takes on an articulate language alien to the quotidian speech circulated in the bar, and in this way he discloses, through metamorphoses and allusions to contemporary Irish nationalist movements and propaganda, the violence implicated in the institution of marriage, while he puts the Irish questions of law, patriarchy and language "at the bar."

Both Fritz Senn and Brigitte Sandquist have signaled the Ovidian metamorphoses that take place within the catalogue, transforming trees into women, the unnamed into names. Senn first observes that "the twenty-nine wedding guests ... have common arboreal stock, but different status (some seem to be engaged in social climbing) and linguistic genealogy, all seem to join gleefully in the Cyclopian trickery of naming, a fluctuation between nouns and names"[101]. Elaborating on Senn's earlier observations, Sandquist maintains:

> In Ovid, specific women become generalized into an undifferentiated class of trees; in Joyce's catalogue, this process is inverted — trees become personified as women. In Joyce's catalogue, common nouns for trees become proper names. A "class" of natural objects (trees) becomes a social class, more particularly a class that designates a highly specific social landscape. The "fir" becomes "Miss Fir Conifer of Pine Valley," and a similar thing happens with such trees as the elm, birch, ash, and olive. The transformation in Ovid's text is toward the natural and organic; in Joyce's text, the natural and organic enter a system of language and discourse. In fact, Joyce's catalogue goes even further than changing trees into women; more specifically, the trees become *names*.[102]

What happens in this catalogue, as Sandquist sees it, is the humanizing transformation of trees.[103] She sees Joyce's human-tree metamorphoses as the inversion of the Ovidian one, retaining the Ovidian violence of transformation with the implicit violence in the very act of naming. She contends that, whereas these tree-women are listed in a highly-structured social space of patriarchy, the naming and titling of the trees designates "an elaborate root system of kinship"[104], further evidenced by the fact that Miss Fir Conifer, our arboreal bride, is given away by her father, the M'Conifer of the Glands. "What Joyce seems to be doing in his metamorphosis of Ovid," Sandquist concludes, is "highlighting the structures and implications of names, genealogical trees, and kinship systems"[105].

Whereas I do not intend to refute the Ovidian parallel in the metamorphoses taking place in this catalogue, it might be of help to locate the transformation in another context -- the Viconian one. From the myth of Apollo and Daphne[106], Vico's *The New Science* interestingly registers the same tree-genealogy comparison:

Apollo begins this history by pursuing Daphne, a vagabond maiden wandering through the forests (in the nefarious life); and she, by the aid she besought of the gods (whose auspices were necessary for solemn nuptials), on standing still is changed to a laurel (a plant which is ever green in its certain and acknowledged offspring), in the same sense in which the Latins use *stipites* for the stocks of families; and the recourse of barbarism brought back the same heroic phraseology, for they call genealogies trees, and the founders they call stocks or stems, and the descendants branches, and the families lineages. (*NS* 531)

For both Vico and Sandquist, therefore, human lineages, or genealogical trees, trace back to the classical folklores of mythology that transform maidens into trees. Slightly different from Sandquist's assertion that Joyce averts the Ovidian metamorphoses by transforming women into trees in his tree wedding catalogue, I view the transfiguring process in the catalogue as a cyclical one. Sandquist puts it rightly when she argues that the trees are, in this catalogue, changed into women, and even women with names and titles according to social proprieties. However, as I have previously elaborated, Joyce is playing with the pun when he composes the tree-run of the catalogue. The tree list, on the one hand, echoes the contemporary afforestation scheme, yet on the other, it implicates the more radical nationalist parades of the INF and the Cumann na mBan. Sandquist's argument persuasively supports the first denotation if we read the catalogue's tree-run as a timber-run in light of the environmental crisis in early-twentieth-century Ireland. However, if we consider another level of meaning to the tree-run, one that involves the political movement of the Irish nationalists, Sandquist's theory would not work for the woodkerne-run engaged here. In this light, the tree list in effect entails a cyclical process of transformation: from the humans (the woodkernes, the Foresters, or the female republicans in green costumes), to the trees, and then to the humans (the personified arboreal guests) again.

The kinship system of feudalism, which Sandquist identifies to be implicated in the social column of this catalogue, reminds us of an ancient Celtic social order of classification: Brehon Law. In Brehon Law, trees were stratified into four classes according to their economic value, with seven species in each class. The highest class is called *Airig Fed* — literally nobles or chieftains of woods or trees, which is composed of oak, hazel, holly, yew, ash, pine, and wild apple. The second class from the top, *Aithig Fedo* (commoners or common trees), includes alder, willow, hawthorn, rowan, birch, elm, and another one, possibly wild berry. In the third class, namely *Fodla Fedo,* meaning the lower orders, there are blackthorn or sloe-bush, elder or bore-tree, white hazel, spindle tree, aspen, arbutus and *crann fir,* possibly juniper. Bushes such as bracken, bog-myrtle, furze or whin, brambles, heather, broom, gooseberry (or wild rose), and ivy all belong to the lowest class of *Losa Fedo,* identified as slave-trees[107]. Whereas in principle there are seven trees in each of the four classes, thus making it twenty-eight in theory, the tree list actually contains twenty-nine trees, with an extra one, *losa fedo* (ivy), included in the class of slaves. To account for this addition, Eoin Neeson indicates the Celtic mystic preference for certain numbers such as three, five, seven, and nine. He informs us that, for instance, the Celtic fighting units "commonly consisted of eight men plus a leader (nine) banded together into a larger unit of three of these plus another leader, twenty-eight, divisible by seven"[108]. In principle, the Celtic tree list is also divisible by seven, then four times seven, with an extra addition to make the overall number twenty-nine.

With the number in mind, the twenty-nine arboreal guests at the tree wedding appear like another Celtic tree list. It is worthy of notice that in the *Little Review* installment where the tree wedding passage was first published in December 1919, there was no indication of a gigantic guest list. Only the wedding couple, along with the two maids of honor, Miss Larch Conifer and Miss Spruce Conifer, were present at the wedding. Even the father of the bride, M'Conifer of the Glands, did not make his appearance in the *Little Review* installment. In other words, Joyce didn't compose the guest list until after the end of 1919. The textual notes to Hans Walter Gabler's 1986 editorial *Ulysses* indicate that the guest list was not added until placard stages three and four (*U-G* 704). In other words, instead of his method of ramification for many of his gigantic lists in "Cyclops," it is very likely that he added the arboreal guest names altogether into this passage within a fairly short period in October 1921[109]. This hypothesis helps us to identify the entire guest list as a whole, and its wholeness, given its numeric correspondence to the twenty-nine trees in the Celtic tree list and its stratified class structure echoing those outlined in Brehon Law, ratifies the guest list's suggestive reference to the ancient Celtic law. Although the tree names in the guest list do not perfectly concur with the stratified arboreal species in the old Irish tree list, many of the trees listed in Brehon Law, such as oak, hazel, holly, ash, hawthorn, rowan, birch, elm, elder, aspen, bog-myrtle, make their way into Joyce's guest names in their original or transpermuted forms. We have, among others, Mrs Holly Hazeleyes, whose name incorporates both holly and hazel, and the name of Lady Sylvester Elmshade contains both elm and Scots pine, which in Latin is called *Pinus Sylvestris*.

The tree wedding catalogue's resemblance to Brehon Law, however, is by no means confined to its ostensible naming or structural parallel to the old Celtic tree list mentioned above. As the popular legislative system of Irish society before the Ulster Plantationsn, Brehon Law was the legal code of the principal social order in Gaelic Ireland. In the first part of this chapter, where the history of Irish forestry was examined, I have previously remarked on Ireland's detrimental condition due to its woodland resources as a result of the replacement of Brehon Law. The renowned Gaelic clan Hugh Roe O'Donnell, the prototype for M'Conifer of the Glands, was significantly among the last Gaelic clan chieftains in Ireland. After the 1601 siege of Kinsale, which O'Donnell lost, the old Gaelic system collapsed, and Brehon Law was no longer officially in use. In the tree wedding, Joyce makes the fictionalized O'Donnell, one of the last Gaelic patriarchs, give away his daughter to John Wyse Nolan, grand high chief ranger of the Irish National Foresters (therefore representing a contemporary chieftain of Irish nationalism), witnessed by the personified tree list out of the ancient Brehon Law. At the end of the third section of this chapter, I briefly noted the popular female figures of Erin on the INF banners and other nationalist prints. These female icons, usually accompanied by towers, wolfhounds, or Celtic harps, are representative of a pure ancient "Irish" Ireland, to which Irish nationalists and Revivalists aspire. The figure of Miss Fir Conifer, on the one hand, alludes to the female nationalists exemplified by the members of Cumann na mBan, and on the other, is the representative Erin on the nationalist banners and prints. In this sense, Miss Fir Conifer (in the ideal image of Erin), given away by her father (the Gaelic chieftain who lost the ancient Celtic order of Brehon Law to the English), is married to the INF high chief ranger John Wyse Nolan, whose enthusiasm for the revival of Irish culture is reflected in his mock-Gaelicised name. In fact, Joyce might have made a wordplay with the INF society here not simply in its Forester pun and its nationalist sentiment but also in its benevolent connection. Vico believes that marriage appeared as the first kind of friendship in the world, claiming that "the true natural friendship is matrimony" (*NS* 260). It is, in Hofheinz's words, "a principle of connubial 'friendship' within a patriarchal framework"[110]. Surely we are not to forget that the Irish National Foresters, though its public performance tends to be outwardly nationalistic, is functionally and fundamentally a friendly society. Sensitive as Joyce is to the institution of marriage as a form of patriarchal oppression, he is very likely to have paid attention to Vico's theory on the marriage-friendship assimilation. Above all, Joyce never loses his sarcastic humor. Henry Flower, Bloom's alternative other, becomes "Senhor Enrique Flor" (*U* 12.1288) who presided at the organ during the tree wedding. Like the mock-Gaelicised name of the bridegroom, his name, too, is ironically transformed into a Portuguese one (echoing John Wyse Nolan's notion: "As treeless as Portugal we'll be soon …" [*U* 12.1258]). And the title Senhor, meaning Lord or Sir, was originally designated a feudal Lord or Sire.

James Connolly regards the two most significant "institutions," class divisions and private property, as England's most devastating colonial imposition on Brehon law as well as on "the communal structures of Gaelic civilization" in

Ireland[111]. He singles out the problems of class structure and property ownership as the "'neo-colonial' condition of Irish political culture" under British colonialism[112]. Similar to Connolly's argument, Irish nationalists and Revivalists look back on an ancient Gaelic culture long before the English Plantations took place, the period in which the Brehon Law was still in practice. The tree wedding in "Cyclops" exposes such contemporary Irish anxiety over the conflict between the old law of feudalism and the new nationalist endeavor to recover the former by "marrying" the much-desired past. Whereas the wedding as a matrimonial ceremony to unite man and woman is, for Joyce, an alternative form of patriarchal oppression, the tree wedding unexceptionally offers the same violence to the patriarchal feudal laws. The personified tree guests from the ancient Celtic tree list in effect not so much witness the wedding ceremony than they "grace" it, thus replacing the colonial obedience to British order with a nationalistic one to the old Celtic order. For Joyce, Irish nationalist fervor for an ideal Irish Ireland, a derivative of past glory, is as ridiculous as marrying an iconic Erin under the law of the fathers. The metamorphoses from ancient law to human to trees and then to human again testifies to Joyce's use of Vico's cyclical historiography and the latter writer's view on the institutions from marriages to burials, and from forests to civilization. Whether it is the trees, the woodkernes, or the women that are being metamorphosed, Joyce's tree wedding ceremony in "Cyclops" dramatizes the violence of the patriarchal institution.

VI.

> *The war is in words and the wood is the world. Maply me, willowy we, hickory he and yew yourselves.*
>
> — James Joyce, *Finnegans Wake*, 98.34–6

> *Mystery is no longer feared, as the great mystery in whose midst we are islanded was feared by those to whom that unknown sea was only a great void. We are coming closer to nature, as we seem to shrink from it with something of horror, disdaining to catalogue the trees of the forest. And as we brush aside the accidents of daily life, in which men and women imagine that they are alone touching reality, we come closer to humanity, to everything in humanity that may have begun before the world and may outlast it.*
>
> — Arthur Symons, *The Symbolist Movement in Literature*, p. 5

There is something contrary to Joyce's writing in Sandquist's argument when she contends that "both the Tree Catalogue and the Citizen's words are *just* words"[113].

Joyce's tree catalogue is by no means "*just* words." As I have elaborated at length throughout this chapter, the historical context of Irish afforestation, the thematic wordplay on the nationalistic Irish National Foresters and contemporary Irish politics, the INF society's performative parades, the Ovidian or Viconian metamorphoses, its reference to Cyclopean gigantism and Vico's philological historiography, and its implicit patriarchal violence of matrimony — all these issues and techniques interplay within this condensed passage on the tree wedding. If the tree catalogue is *just* words, then it might indeed be, like the wood to Joyce, *just* the world in which there are wars and words and war in words. That, in other words, makes the world, and this is the world where Vico's civilization began: from the Cyclopean myth, the forests of the giants.

Emer Nolan specifically highlights "the historical context of issues about law, history and government which is at stake" in "Cyclops"[114]. In the Cyclopean eye of this episode where the forest becomes tree-run becomes woodkerne-run, the law of the fathers is transfigured into contemporary human forms — figures of personified patriarchal law. The forest, for both Vico and Joyce, is not so much the environmental resource to be preserved as the product of human appropriations throughout the history of civilization and colonialism. Joyce's environmental attitude toward the woodlands is more political, cultural, and socio-historical than it is ecological. The metamorphoses taking place within this catalogue in fact confirm, rather than defy, the distance between human and nature. Vico's imaginative metaphysics shows that "man becomes all things by *not* understanding them" because "when he does not understand he makes the things out of himself and becomes them by transforming himself into them" (*NS* 405). Becoming is therefore an appropriating act, an act too impossible to be negotiated when nature is incomprehensible. And in "Cyclops" the act of transformation — with the woods becoming tree-run becoming woodkerne-run becoming old Celtic law becoming tree-women guests — becomes, like the linguistic incommunicability of Myles Joyce in "Ireland at the Bar," the manifestation of an impossible reconciliation between the history of nature and that of civilization. Behind such a process of metamorphosis is the Viconian historiography which grants new meanings to the historiography of nature, civilization, language, and ecosystem. It poses questions on the politics of performative demonstrations of nature, its colonial and nationalistic undertones, and challenges the boundary between civilization and nature, and that between language and ecosystem, with a philological historiography that sees history and language not only as mutually transmissible but also cyclically and ecologically metamorphosis-able.

Joyce's Viconian road in *Finnegans Wake* goes round and round to where the book begins: "riverrun" (*FW* 3.1). The running of a river, as Rosa Maria

Bosinelli indicates, "contains overtones of the Italian word 'riverrun,' that is, 'they'll come again'"[115]. Like the second coming of the "riverrun," maybe the "tree run," the demystified "catalogue [of] the trees of the forest"[116] in "Cyclops," also anticipates the second coming of the forest, which is the beginning of civilisation, and would, though rather ironically, see to "A Nation Once Again" (*U* 12.891; 12.916–7).

Notes

1. An earlier version of some parts of this chapter speared in Eco-Joyce: The Environmenal Imagination of James Joyce (Cork: Cork University Press, 2014), under the title "The Tree Wedding and the (Eco)Politics of Irish Forestry in 'Cyclops': History, Language and the Viconian Politics of the Forest." The current chapter in this book has since then been fully expanded and contains other dimentions of cultural analysis of politics of the forest in *Ulysses*. I am grateful for Cork University Press for permitting me to reprint the previously published materials in this chapter.
2. The Mesolithic people who arrived in Ireland between eight and ten thousand years ago used to clear the forests "by felling, by deliberate fire and by grazing stock" (Neeson 1997:134).
3. Joyce referred to it as "this disease-ridden swamp" for whose failed reforestation the English government was to blame (*OCPW* 144).
4. "The Land Act of 1881 enabled land to be transferred from landlord to tenant, complete with standing timber. Both new owners and vendors found it convenient to sell plantations for additional cash – […] the landlords selling for additional profit before the sale of land, new owners selling after purchase to recoup some of their outlay. With the passing of the Land Act of 1903 transfers took place on a greater scale than ever and much timber was sold […]" (Neeson 1997:150).
5. Afforestation involves planting seeds or saplings to make a forest on land which has not recently been forested, or which has never been a forest, such as bogland.
6. Eoin Neeson, *A History of Irish Forestry* (Dublin: Lilliput Press, 1991), p. 5.
7. A. C. Forbes, "Tree Planting in Ireland during Four Centuries," *Proceedings of the Royal Irish Academy* (1932–1934), p. 168.
8. Eileen McCracken, *The Irish Woods Since Tudor Times: Distribution and Exploitation* (Newton; Abbott: David & Charles, 1971), p. 15.
9. Eoin Neeson, "Woodland in History and Culture" in *Nature in Ireland: A Scientific and Cultural History*, ed. by John Wilson Foster (Dublin: Lilliput Press, 1997), p. 140.
10. Joyce, *Occasional, Critical, and Political Writing*, p. 198.
11. McCracken, p. 98.

12. Gifford, p. 355.
13. Ibid.
14. Ibid., p. 356.
15. Neeson, "Woodland in History and Culture," p. 142.
16. Qtd. in Neeson, "Woodland in History and Culture," p. 142.
17. Neeson, *A History of Irish Forestry* (Dublin: Lilliput Press, 1991), p. 62-3.
18. Ibid., p. 67.
19. See *Report of the Departmental Committee on Irish Forestry* by DATI (Dublin: HMSO, 1908), p. 35.
20. Neeson, "Woodland in History and Culture," p. 149.
21. Ibid.
22. Forbes, "Tree Planting in Ireland during Four Centuries," p. 191.
23. See DATI, *Report of the Departmental Committee on Irish Forestry*, pp. 45-6.
24. Ibid., p. 32, my italics.
25. Ibid., pp. 6-7.
26. Ibid., p. 6.
27. Ibid., p. 11.
28. Qtd. in the DATI Report, p. 15.
29. DATI, *Report of the Departmental Committee on Irish Forestry*, p. 15.
30. Neeson, *A History of Irish Forestry*, p. 132.
31. With the information provided, we can gather the speaker is Joseph Nolan, who served as representative for South Louth at the Parliament. Source of information: http://www.jbhall.freeservers.com/a_brief_history.htm (1886 County Louth: a Brief History.)
    Another possible source for the "Mr Nolan" of "Cyclops" could also be Colonel John P. Nolan, M.P., who was among the 14 members of the Agricultural Board of the Department of Agriculture and Technical Instructions for Ireland, listed in 1904 *Thom's Directory*, p. 836.
32. Gifford, p. 351.
33. Gifford, pp. 341, 351.
34. Gibson, *Joyce's Revenge*, p. 24. In *Joyce's Revenge*, Gibson brilliantly analyses the influence of Balfour's Conservative-Unionist government (1895-1905) on Joyce's writing of *Ulysses* (on "Telemachus" in particular). See pp. 21-41.
35. For some exemplary references for Joyce's interest in Griffith, see Richard Ellmann's *The Consciousness of Joyce*, pp. 86-90; Manganiello's *Joyce's Politics*; Fairhall's *James Joyce and the Question of History*, pp. 96-97, 173-77; Andrew Gibson's *Joyce's Revenge*, pp. 49-59, 119-26, 193-4, among others.
36. Richard Ellmann, *The Consciousness of Joyce* (London: Faber and Faber, 1977), p. 87.
37. Manganiello, *Joyce's Politics*, p. 118.
38. Ibid.
39. Ibid., pp. 119-126.

40. Gibson, *Joyce's Revenge*, pp. 119–123.
41. Younger p. 78, qtd. in Gibson, *Joyce's Revenge*, p. 123.
42. Arthur Griffith, *The Resurrection of Hungary: A Parallel for Ireland* (1904), with introduction by Patrick Murray (Dublin: University College Dublin Press, 2003), p. 154.
43. Ibid., p. 155.
44. Ibid.
45. Ibid.
46. Manganiello, *Joyce's Politics,* p. 124.
47. Griffith, p. 155.
48. Gibson, *Joyce's* Revenge, p. 119.
49. Ibid., p. 122.
50. Ibid., p. 115.
51. Ibid., p. 108.
52. Ibid., p. 115.
53. Ibid., pp. 114–5.
54. The Right Honorable Joseph Hutchinson, who was the Lord Mayor of Dublin in 1904, not only co-founded the INF but also had been enthusiastically involved in its activities. In 1904 he also served as the general secretary to the Society and appeared in Forester events across the country.
55. Source of information from the website of Ancient Order of Foresters Friendly Society: http://www.forestersfriendlysociety.co.uk/our-history.aspx (Our History)
56. John Francis Campbell, *Friendly Societies in Ireland 1850-1960: with particular reference to Ancient Order of Hibernians and the Irish National Foresters*, unpublished M. Litt. Thesis, Trinity College Dublin, October 1998, pp. 54–55.
57. This correspondence, however, by no means suggests Bloom has any affiliation with the Irish National Foresters, since the society intends to be comprised only "of Irishmen by birth or descent" (21 Dec 1907, *Irish Independent*).
58. Anthony D Buckley, "'On the club': Friendly Societies in Ireland," in *Irish Economic and Social History* 14 (1987), pp. 56-7.
59. Campbell, p. 97.
60. Captain H. B. C. Pollard, *The Secret Societies of Ireland: Their Rise and Progress* (London: Philip Allan & Co., 1922), p. 275.
61. Neil Jarman, *Material Conflicts: Parades and Visual Displays in Northern Ireland* (Oxford: Berge, 1997), pp. 141–2.
62. Ibid., p. 142.
63. Campbell, p. 87.
64. Campbell, p. 87.
65. Fritz Senn, "'Trivia Ulysseana' IV" in *James Joyce Quarterly* 19.2 (Winter 1982), p. 169.
66. Jarman, p. 142.

67. Ibid., p. 123.
68. Campbell, p. 237.
69. Emer Nolan, *James Joyce and Nationalism* (London: Routledge, 1995), p. 51.
70. Ibid.
71. C. P. Curran, *Under the Receding Wave* (London: Gill and Macmillan, 1970), p. 26.
72. Ibid.
73. Ibid.
74. Ibid.
75. Jarman, p. 163.
76. Campbell, p. 220.
77. Ibid., p. 225.
78. It has been located by critics including Bernard Benstock who assume that Joyce bases John Wyse Nolan on John Wyse Power (1859–1926), a renowned nationalist and journalist at the time.
79. Owen McGee, "Power, John Wyse," *Dictionary of Irish Biography*. Cambridge: Cambridge University Press, 2009. <http://dib.cambridge.org/viewReadPage.do?articleId=a7469>
80. William Murphy and Lesa Ní Mhunghaile, "Power, Jennie Wyse," *Dictionary of Irish Biography*, Cambridge: Cambridge University Press, 2009. <http://dib.cambridge.org/viewReadPage.do?articleId=a7454>
81. Douglas Hyde, *Beside the Fire: a Collection of Irish Gaelic Folk Stories* (1910), p. xxvii.
82. Ewan Morris, *Our Own Devices: National Symbols and Political Conflict in Twentieth-Century Ireland* (Dublin: Irish Academic Press, 2005), pp. 12–25.
83. Fritz Senn, "Ovidian Roots of Gigantism in Joyce's 'Ulysses,'" in *Journal of Modern Literature* 14.4 (Spring 1989), p. 567n.
84. Brigitte L. Sandquist, "The Tree Wedding in 'Cyclops' and the Ramifications of Cata-logic," in *James Joyce Quarterly* 33.2 (Winter 1996), p. 201.
85. Bernard Benstock, "Vico… Joyce. Triv.. Quad" in *Vico and Joyce*, Donald Phillip Verene, ed. (Buffalo: SUNY Press, 1987), p. 66.
86. Max Harold Fisch and Thomas Goddard Bergin, "Introduction" in *The Autobiography of Giambattista Vico*, trans. by Max Harold Fisch and Thomas Goddard Bergin (Ithaca: Cornell University Press, 1944), p. 97.
87. A. Walton Litz, "Vico and Joyce," in *Giambattista Vico: an International Symposium* (Baltimore: the Johns Hopkins Press, 1969), p. 249.
88. Stuart Gilbert, *James Joyce's 'Ulysses': A study* (London: Penguin, 1963), p. 338.
89. Samuel Beckett, "Dante… Bruno. Vico… Joyce" in *Our Exagmination Round His Factification for Incamination of Work in Progress* (New York: New Directions, 1962), p. 4.
90. Gilbert, p. 46.
91. Beckett, p. 5.

92. Joseph Mali, "Mythology and Counter-history: the New Critical Art of Vico and Joyce" in *Vico and Joyce*, Donald Phillip Verene, ed. (Buffalo: SUNY Press, 1987), p. 40.
93. Rosa Maria Bosinelli, "'I use his cycles as a trellis': Joyce's Treatment of Vico in *Finnegans Wake*" in *Vico and Joyce* (Buffalo: SUNY Press, 1987), p. 125.
94. Sandquist, p. 201.
95. Thomas C. Hofheinz, *Joyce and the Invention of Irish History*: Finnegans Wake *in context* (Cambridge University Press, 1995), pp. 150–55.
96. Ibid., p. 141.
97. Ibid.
98. Litz, p. 254.
99. Robert Spoo, *James Joyce and the Language of History: Dedalus's Nightmare* (Oxford: Oxford University Press, 1994), p. 84.
100. Ibid.
101. Senn, "'Trivia Ulysseana' IV," p. 167.
102. Sandquist, p. 198.
103. However, what happens in the catalogue is not, as Sandquist argues a few lines earlier, the continuation of maiden-tree transformation that starts as early as the beginning of "Cyclops," where "lovely maidens sit in proximity to the roots of the lovely trees singing the most lovely songs ..." (*U* 12.78–80). Whereas these maidens are indeed illustrated in the mock-bardic language of a classical epic, they show no sign of the slightest transformation. The human-tree metamorphoses does take place — yet, in my view, in "Cyclops" it takes place in the tree catalogue only.
104. Sandquist, p. 199.
105. Ibid.
106. Interestingly, Miss Daphne Bays (bays being part of the laurel family) appears among the tree wedding guest in the list in "Cyclops" (*U* 12.1270).
107. See Fergus Kelly, "The Old Irish Tree List," 108–24.
108. Neeson, *A History of Irish Forestry*, p. 29.
109. Michael Groden, "'Cyclops' in Progress, 1919" in *JJQ*, Vol. 12, No. 1/2, Textual Studies Issue (Fall 1974 – Winter 1975), p. 133.
110. Hofheinz, p. 146.
111. See Gibson, *Joyce's Revenge*, p. 96n.
112. Ibid., p. 96.
113. Sandquist, p. 203.
114. Nolan, p. 112.
115. Bosinelli, p. 125.
116. Arthur Symons, *The Symbolist Movement in Literature* (Kessinger Publishing, 2004), p. 5.

4 PASTORAL
Social Languages, Politics, Nature and Nation

Emer Nolan puts it rightly when she mentions how "the violence signified by 'Cyclops' as a whole is not just the crude words and physical force of the citizen, but also the violent clashes it demonstrates between different languages"[1]. Here, by "different languages," Nolan does not refer to national languages *per se*, but the textual languages or social discourses employed by different levels of narrative. Language is undoubtedly, as exemplified throughout *Ulysses*, a structural construction Joyce forces his readers to meditate on. In the chapter of "Cyclops," the narrative flow keeps getting interrupted not so much by streams of consciousness as by blocks of catalogues. These diversified intercessory catalogues, however, offer alternative narratives of the nation that justly demonstrate the problematics of language when politics[2] and the question of the nation are put under scrutiny.

The simultaneous alternativeness of diverse national discourses in "Cyclops" results in a mosaic of styles on the rhetorical level, like the "various forms and degrees of *parodic stylization* of incorporated languages"[3] that, according to Bakhtin, predominate the form of novels. "Incorporated languages" like these, as a result of the interpolation of catalogues, stratify the episode of "Cyclops" into multiple layers of linguistic and social discourses, and enable "dialogical" interrelationships between different "utterances" and languages to converse with each other. This chapter strives for an exclusive analysis of the different levels of the narrative, or "language stratification", that the first (or first-emerged) catalogue in "Cyclops" has to offer. It begins with my reading of the "Inisfail the fair" paragraph as an ideal pastoral passage alluding to James Clarence Mangan's poetic translation of "Prince Aldfrid's Itinerary through Ireland," and ponder upon the passage's rhetorical complexity that involves pastoral literary tradition and Manganesque scholarship. In the second part of this chapter, I refer to Michael Groden's genetic studies on "Cyclops" and venture to examine the "Inisfail the fair" passage alongside Joyce's writing process, reading the passage as Joyce's response to Anglo-Irish Revivalist writings together with the episode's "Aeolus" connection. The juxtaposed reading of "Aelous" and "Cyclops" brings out traces of interconnection between the two episodes with regard to political rhetoric and nationalism, as well as mass culture and sociopolitical background, issues I continue to investigate in the third section of this chapter. Lastly, I proceed with discussions on the social geography of the Dublin Corporation Market, where the "Inisfail the fair" catalogue is set. Inspired by the Spenserian parallel to the passage's linguistic structure, I raise the question concerning coloniality and spatiality, and by specifically focusing on the catalogue, endeavor to contemplate the diverse aspects of politics this passage enacts. Overall, by looking at idyllic passages from "Cyclops" and "Aeolus" through different layers of languages and the socio-politics of the cultural production of the pastoral, this chapter hopes to reconsider the idyllic representations in the context of nationalism and consumer culture in fin-de-siècle Ireland, as well as the eco-politics of the literary imagination of the pastoral in *Ulysses*.

I.

The episode of "Cyclops" encloses an interestingly extensive amount of passages that appear to depart from its overt political context. One of the first few catalogues interweaves a picturesque narrative that impresses us with its Romantic pastoral language: a "pleasant land it is in sooth of murmuring waters, fishful streams where sport" different kinds of fishes "too numerous to be enumerated"; "[i]n the mild breezes of the west and of the east the lofty trees wave in different directions"; and "[l]ovely maidens sit[ting] in close proximity to the roots of the lovely trees singing the most lovely songs while they play with all kinds of lovely objects" when "heroes voyage from afar to woo them" (*U* 12.68–86). This innocent, rather idyllic language reminds one of Christopher Marlowe's passionate shepherd when he chants: "[t]here will we sit upon the rocks/And see the shepherds feed their flocks,/By shallow rivers, to whose falls/Melodious birds sing madrigals. [...]"[4]. And yet, the carefree pastoral scene functions beyond the innocent poetics of shepherd's love or transcendental nature; it furthermore "remove[s] the location of idyll from the tumult of everyday life into the simple pastoral state and assign[s] its period before the *beginning of civilisation* in the childlike age of man"[5]. In this sense, pastoral is by no means exclusively naive or simple; it is, rather, a literature with hidden contradictions for post-industrialized modern society.

Roger Sales defines pastoral as "essentially escapist in seeking refuge in the country and often also in the past; that it is a selective 'reflection' on past country life in which old settled values are 'rescued' by the text; and that all this functions as a simplified 'reconstruction' of what is, in fact, a more complex reality"[6]. Raymond Williams' definition holds a similar view except for his more emphatic opposition between the country and the city, as well as the past and the present. In *The Country and the City*, he notes how in pastorals "an ordered and happier past [is] set against the disturbance and disorder of the present. An idealisation, based on a temporary situation and on a deep desire for stability, served to evade the actual and bitter contradictions of the time"[7]. Whether to reconstruct social complexity or to accommodate escapism from disorder, pastorals draw attention to disruptive contrasts — "the spatial distinction of town (frenetic, corrupt, impersonal) and country (peaceful, abundant), and the temporal distinction of past (idyllic) and present ('fallen')"[8]. In this sense, pastoral imaginations function, by envisioning a natural world beyond reality, as nostalgic responses to evade as well as to relieve contradictory modern anxieties.

Opening with "In Inisfail the fair there lies a land" (U 12.68), Joyce's pastoral catalogue is an idyllic rewriting of James Clarence Mangan's translation of "Prince Aldfrid's Itinerary through Ireland," a medieval Irish poem attributed to Aldfrid, who is a Northumbrian king in the seventh century[9]. The poem is a recounting of Aldfrid's journey to Ireland, in which he finds "[a]bundant apparel and food for all," "[g]olden and silver [he] found in money;/Plenty of wheat and plenty of honey;/[...]Found many a feast, and many a city [...]"[10]. He tells how, in respective areas of Ireland, he finds "the good lay monks and brothers," "[k]ings and queens and poets a many," "[R]iches, milk in lavish abundance," "the glorious, bravest heroes, ever victorious," "[h]ardy warriors, resolute men," "[f]lourishing pastures, valor, health,/Song-loving worthies, commerce,/wealth," and "[s]weet fruits, good laws for all and each,/Great chess-players, men of truthful speech, [...]"[11]. In brief, the poem records the glorious past of Ireland during a medieval Golden Age. With the national glory the poem evokes, Joyce pastiches the pastoral form and inserts it into the beginning few paragraphs of "Cyclops," from which emerge gigantic questions on nature, nation, and politics.

The innocent picturesque catalogue of Inisfail the fair is, accordingly, interwoven with different narrative schemes: from the outset, the Homeric design[12]; from the language, the Romantic pastoral tradition; from the annotation, "Aldfrid's Itinerary"[13]. Whereas Homer provides a blueprint for the narrative structure, "Prince Aldfrid's Itinerary through Ireland" intermingles with pastoral writing and reminiscences about the heroic, prosperous, and abundant past of Ireland. Under the pastoral picturesqueness lay contradictions between the urban and the rural, the present and the past. Such contradictions call into question the post-Famine Irish condition in the wake of Famine deaths and emigration: "[b]y 1850, Ireland had one of the most commercially advanced agricultures in the world, [...] was developing one of the world's densest railway systems[, and] also contained the fifth greatest industrial city on earth [...]"[14]. Yet, as Terry Eagleton informs us, "The country as a whole had not leapt at a bound from tradition to modernity" due to its "combined and uneven development."[15] [16] The problem of uneven development in Ireland, with the economic and developmental gap between city and country, results in urban anxiety for the underdeveloped rural areas, economic exploitation of agrarian industry, and drastic differences in the ideologies of the city and the country. Although Eagleton contends that "the naturalizing strategies of English ideology [may] not stick so well in Ireland"[17], it is impossible to consider Irish political and social contradictions without pondering on the natural, or ecocentric, politics of the nation.

The conjunction between the disjunctive social problems implicated in the pastoral and the nostalgic national glory projected in "Prince Aldfrid's Itinerary through Ireland" further complicates the interwoven complexity of the catalogue. If the pastoral works as a "greening device" to conceal the anxieties derived from national nostalgia and uneven civil development, "Prince Aldfrid's Itinerary through Ireland," in celebrating Ireland's glorious past, is framed within the pastoral tradition to reflect the differentials between Ireland's past and present, its country and city. The original Irish text is attributed to an Anglo-Saxon king of Northumbria, and the Anglophone version of the poem Joyce employs is a translation by James Clarence Mangan. The authorial narrative in "Cyclops," therefore, consists of multiple layers of retelling, in which Joyce retells the pastoral retold by Mangan based on an Anglo-Saxon king's itinerary record.

Sean Ryder observes that, in Mangan's poetic translations from Irish, there exists "[a] [...] paradoxical relation [...], insofar as the Gaelic tradition is part of his heritage (his father in fact would almost certainly have been an Irish speaker), yet Irish for Mangan is a foreign tongue that he can only encounter indirectly through the literal translations of his friends"[18]. This circumstance was similar to the provincial linguistic application of English among Irish writers, whose relationship to English was more instrumental than historical or cultural. "In one critical passage," as Ryder points out, "Mangan expresses his love for 'our own mother tongue', by which he means English; yet his obsessive verbal play, his incessant punning, are all symptomatic of an insecure relation to English, albeit which, like Joyce, he exploits as a kind of linguistic freedom"[19].

It is perhaps this kind of linguistic freedom that prompts James Joyce to identify with James Clarence Mangan, whom he once praised: "I know of no other piece of English literature where the spirit of revenge has attained such heights of melody" (*OCPW* 134). While he fondly regards Mangan's lyrical technique as "one of the most inspired singers that ever used the lyric form in any country"[20], Joyce also sees Mangan as a poet who fails his "essential effort [...] to liberate himself from the unpropitious influences of such idols which corrupt him from the inside and out" (*OCPW* 135). By this he refers to the national history of Ireland, which "encloses [Mangan] so straitly that even in his moments of high passion he can but barely breach his walls" (*OCPW* 135). Joyce remarks on Mangan as "a miserable, reedy, and feeble figure [, in which] a hysteric nationalism receives its final justification" (*OCPW* 136); yet his admiration for and empathy for him are revealed not only in his Trieste speeches but also in textual references to Mangan's poetry. It is with his recognition of Mangan as a national poet, probably also with the paradoxical linguistic compassion they share, that Joyce incorporates "Prince Aldfrid's Itinerary through Ireland" into a Cyclopean context that is political on the one hand and paradoxical on the other.

Mangan's nationalistic attitude represented in his translation of "Prince Aldfrid's Itinerary through Ireland" remains ambiguous, partly due to his contention that "[w]e should never judge of authors from their works"[21] and regardless of his later earnest participation in the patriotic group Young Ireland. Yet, however Mangan regards himself in the nationalistic literary movement, "nationalism was inevitably part of the cultural air Mangan breathed, and his poems are an interesting reflection of the changing emphases and issues within Irish nationalism over the course of twenty years"[22]. Mangan's "Prince Aldfrid's Itinerary through Ireland," considered by critics as "very close to the original in wording and tone"[23], may not be laden with the same nationalistic implications exemplified in many other poems and poetic translations of his. Nevertheless, his obscure political position in this poem didn't hinder Joyce from taking it as the textual source to be incorporated into its pastoral Cyclopean catalogue in the particularly political chapter of "Cyclops." Ryder's notes to Mangan's poem inform us how, in a later reprinted version by the Irish-American newspaper *An Gaodhal-The Gael* in 1887, the poem was retitled "Ireland under Irish Rule"[24]. The renamed title explicitly contrasts with Irish conditions of the time: it was the time of "Ireland under British rule." Hence the poem is given a nationalistic voice disregarding Mangan's authorial intention or position: the reminiscence about a glorious Irish sovereignty gives a nationalistic nostalgia to the poem, with the intertextual paradox of linguistics and socio-politics in Ireland its catalyst.

Apart from the paradoxical relation between Mangan's English translations of Irish poetry and his anti-British Irish identity,[25] between an ideal national epic and its colonized present, Joyce further complicates the paradoxical use of Mangan's Anglo-Irish-Anglo-Saxon poem by rewriting it in a Romantic idyllic style. The passage's texture of multiple linguistic layers is incorporated into multiple narrative layers to construct an interwoven context of languages, styles, literary traditions, colonial power politics, and paradoxical identities in Ireland. By applying this Cyclopean catalogue interwoven with multiple layers and threads to the paragraph that brings forth an urban, public space — Barney Kiernan's pub — in which politics is not merely pondered but directly confronted, Joyce puts Mangan in a position beyond a national poet. He transforms Mangan's poem, with pastoral elements, linguistic paradox, socio-politics and coloniality, into a signifier more complicated than that of "a romantic, a would-be herald, a prototype for a would-be nation" (*OCPW* 136).

Marjorie Howes argues that "Joyce takes up the issue of narrating the Irish nation in a kind of geographical representation that [...] foreground spatial scales — the local, regional, international"[26] and:

> it is precisely through these alternative scales, and the opportunities and obstacles they pose for imagining the scale of the national, that Joyce's engagement with the problematic of the nation appears most vividly.[27]

In her essay, Howes supports her argument with observations from Joe's handkerchief in "Cyclops," elaborating how the "intricately embroidered ancient Irish facecloth" (*U* 12.1438–39) offers an "alternative narrative of the nation" in a different spatial scale by way of fabricating national nostalgia. This kind of alternative narration can also be found in the pastoral catalogue of "Innisfail the fair". The idyllic picture in this catalogue, interwoven with diverse scales from socio-politics, colonialism, linguistics, to literary tradition, parallels the narrative's inauguration of urban pub culture and modern politics. The complexity of its multiple layers of narratives and languages exemplifies an "alternative narration of the nation" in its multiple layers of reading and representing the nation, whereas Joyce's occasional jokingly-misplaced items within the gigantic lists seem to jest right at the serious heart of Irish politics.

II.

Michael Groden's 1977 book *"Ulysses" in Progress*, one of the pioneering genetic readings of *Ulysses*, traces Joyce's writing process and divides it into three stages: the early stage, characterised by the opening episodes, in which Joyce adopts his renowned third-person stream of consciousness narrative; the middle stage, in which Joyce disregards his previous technique and replaces it with newly innovated ones exemplified by his composition of "Cyclops"; and the last stage, in which his developed encyclopedic style pervades the last few episodes. In the chapter "The Middle Stage: 'Cyclops,'" Groden observes how Joyce came up with a "radically different" episode than the previous ones: it is "a combination of a first-person naturalistic description of the scene in Barney Kiernan's pub, narrated by an unnamed debt collector, and a series of elaborate extensions and exaggerations of incidents in the pub"[28]. Interestingly, Groden further notices that, instead of setting up his plan for "Cyclops" in its final published form — the "complete break with the stream of consciousness technique in the double focus provided by the first-person narrator and the succession of 'gigantic' parodies" — Joyce "created the parodies first, the barroom scene came soon after, and the narrative voice developed last"[29]. The more recently recovered manuscript MSS 36,639/10[30] has, apart from its significance in confirming Joyce's writing process for "Cyclops," given further firm evidence in support of Groden's arguments some three decades earlier.

According to V.A.8 copybook, the earliest extant "Cyclops"[31] manuscript, it is likely that Joyce started drafting the episode with the parodic paragraph of "Inisfail the fair" (V.A.8 p.1r, *U* 12.68–86), and that Mangan's poetic language was not in his initial plan. As critics, Groden among others, point out, the fact that "Cyclops" begins with a gigantic parody signifies Joyce's initial mock-heroic design for the episode. Phillip F. Herring describes that "[t]he original beginning of 'Cyclops' located the reader geographically in Dublin's St. Michan's parish, temporally in a general parody of Irish heroic poetry"[32]. To be more precise on Joyce's use of parody, Andrew Gibson sees the entire episode as Joyce's revenge on the contemporary Revivalist movement "a sustained assault on Anglo-Irish, revivalist historiographies and constructions of Irish history, and the politics and aesthetics implicit in them"[33]. Gibson supports his argument with examples drawn from Anglo-Irish Revivalist writers such as Ferguson, O'Curry, Yeats, Lady Gregory, Douglas Hyde, and above all, Standish O'Grady, thereby claiming Joyce's parodic imitation of these writers' "pseudo-bardic" and "plain-historical" style, in which "Revival versions of the ancient tales" are "moralized, cleansed of various elements originally intrinsic to them: the obscene, the comic, the fantastic, grotesque, monstrous, and frivolous"[34]. If Gibson is correct, Joyce's language in "Cyclops" does not simply function to ridicule the Anglo-Irish Revivalist writings, but also operates to supplement his bardic parodies with "the obscene, the comic, the fantastic, grotesque, monstrous, and frivolous"[35]. In this way, Joyce questions the problematics of historiography (especially those constructed by Anglo-Irish Revivalists) and mockingly recreates one by "filling in" what seems to be "lacking" (in this case, the vulgar and the extravagant language in Barney Kiernan's bar) in these revivalist writings.

The "Inisfail the fair" paragraph at the beginning of V.A.8 copybook reads as follows:

> ~~In green Erin of the west~~ [Inisfail the fair] there lies a land, the land of holy Michan. There rises a watchtower beheld from afar. There sleep the dead as they in life slept, warriors and princes of high renown. There wave the lofty trees of sycamore; the eucalyptus, giver of good shade, is not absent: and in their shadow sit the maidens of that land, the daughters of princes. They [sing and] sport with silvery fishes caught in silken nets; their fair white fingers toss with gems of the [fishful] sea, ruby and purple of Tyre.

> And men come from afar, heroes, the sons of Kings, to woo them for they are beautiful and all of noble stem. (V.A.8, p.1r)

From this obviously shorter version of the passage, we can discover that there were only the merest traces of lists in his first draft,[36] and that Joyce has made few revisions (except later augmentations in catalogues) since this version. Significantly, among the few changes, Joyce crossed out "In green Erin of the west" and replaced it with "Inisfail the fair," which, as Herring notes, is a homage paid to James Clarence Mangan. However, Herring self-contradictorily responds to Weldon Thornton's suggestion of the implied influence of Mangan[37] by commenting that he regards it as a slighter impact than Thornton supposes, since "Joyce's notesheets show that he gathered notes from heterogeneous sources for his parody of the heroic muse"[38].

Without further explanation of how Joyce's heterogeneous notes textually testify to Mangan's "slight" influence on him, Herring's argument sounds implausible to me. Joyce's change of words to "Inisfail the fair" itself crucially marks his specific intention to allude to Mangan, not to mention the fact that, according to William M. Schutte, references to Mangan's poetry recurred a few times in "Cyclops" alone[39]. Gibson's contention regarding Joyce's parodic use of Anglo-Irish revivalist writings here comes into play: if, as Gibson suggests, Joyce imitated the pseudo-bardic style to resist, or even to seek revenge on, the Anglo-Irish revivalist movement, Joyce's allusion to Mangan would have complicated his use of bardic parodies. Hence question arises: does Joyce intentionally contradict himself with his allusion to Mangan, a national poet he so praises, in a parodic passage intended to ridicule the Anglo-Irish Revivalists?

Joyce may have already offered an answer to this paradox in his Trieste lecture on Mangan in 1907. In the speech, he states what he regards as poets' necessary freedom and Mangan's incapability to liberate himself:

> Poetry takes little account of many of the idols of the market-place, the succession of the ages, the spirit of the age, the mission of race. The essential effort of the poet is to liberate himself from the unpropitious influences of such idols which corrupt him from the inside and out, and it would certainly be untrue to assert that Mangan made this effort. *The history of his country encloses him so straitly that even his moments in high passion he can but barely breach its walls.* (*OCPW* 135, italics added)

Such may be the nightmarish nature of history from which Stephen is "trying to awake" (*U* 2.377): Joyce sees the confinement of Ireland's history, specifically its devouring enclosure, in Mangan's tragic inheritance of "the latest and worst part of a tradition upon which no divine hand had drawn out the line of demarcation, a tradition which dissolves and divides against itself as it moves down the cycles"

(*OCPW* 135–6). Mangan is therefore, for Joyce, not so much a pseudo-innocent national poet of the Anglo-Irish revival movement than a "miserable, reedy, and feeble figure" in which "a hysteric nationalism receives its final justification" (*OCPW* 136). He has inevitably become a product of the emerging nationalist movement as a result of his confinement by the nation's history. Inevitably, his voice poses part of the nationalist narrative.

Andrew Gibson notes the "connections binding the emergent nationalist culture to Revival culture" as well as "the latent continuities between their respective views of history" that Joyce is surely aware of[40]. As he further claims, "[t]he relationship between the two cultures partly resembled that between Revival culture and English culture. The demands were repeated for purity. The reality, again and again — as Joyce underlines — was interinvolvement"[41]. What Gibson points out as present in Joyce's time is the simultaneity and interaction of Irish nationalism, Anglo-Irish revival culture, and English cultural imperialism. Insofar as we consider Joyce's political schemata for "Cyclops," his "interinvolvement" that incorporates bardic parodies and Manganesque allusions becomes hardly surprising, though no less paradoxical.

Joyce's choice of the Manganesque allusion, instead of the "green Erin of the west" he initially composed, nonetheless does not signify Joyce's disposal of his original design. Instead, I surmise that, during his later revision, he integrates the replaced phrase "green Erin of the west" with another line from the copybook, "their fair white fingers toss the *gems of the sea*" (V.A.8, p. 1r, my italics), which Groden assumes Joyce drops[42], and places it among the "Aeolus" headlines, in the title "ERIN, GREEN GEM OF THE SILVER SEA" (*U* 7.236). The similarity of the wording is by no means coincidental. Groden informs us that Joyce added the headlines in "Aeolus" probably around mid-September 1921; in other words, when he started writing "Cyclops" in 1919, and in "Aeolus"'s first appearance in the *Little Review* in October 1918, the headlines had not yet emerged. More clearly, Groden's comparison of Aeolian subheads appeared in the first *placards* and those as printed in the published book provides evidence to my assumption: instead of "Erin, Green Gem of the Silver Sea" as we see it, the initial subhead Joyce devises is "Erin, the Gem of the Sea"[43], which, as I mentioned, is the exact phrase Joyce deleted from his original passage in the V.A.8 copybook. The Cyclopean influence on this subhead is consequently beyond doubt.[44]

Groden indicates "a certain amount of cross-exchange" between the later revision of "Aeolus" and the already written later episodes incurred due to Joyce's working time frame; however, among the couple of examples Groden mentions[45], he doesn't specify the revised "Aeolus"' cross-exchange with "Cyclops." In his later chapter on the middle stage, he marks the link between "Cyclops" and

"Aeolus" regarding each's character list when he discusses Joyce's initial design for "Cyclops":

> Bloom's rationale for going to Barney Kiernan's existed from the start, although Joyce does not mention it in draft V.A.8: Bloom has agreed to meet Martin Cunningham there to arrange for the payment of Paddy Dignam's insurance (313.4–9). The preparation for this in "Sirens" (280.29–30) occurs in the fair copy of that episode (fol. 31); hence, it existed before Joyce began to write "Cyclops." However, he originally planned explicit links with other episodes that do not exist in published versions. In the copybook, the patrons of the pub include Stephen Dedalus, Professor MacHugh, and O'Madden Burke. Along with Lenehan, J. J. O'Molly, and Ned Lambert, who characteristically remain in the pub throughout all the revisions, they constitute the central group of characters in "Aeolus." As a further parallel, Myles Crawford, chauvinistic editor, is replaced by Cusack, chauvinistic citizen.[46]

Groden concludes with his observation that "the original close links between the two episodes are not surprising, since both deal more explicitly with political and social themes than do any other episodes"[47]. Despite the fact that I agree with Groden's concluding contention, I find in "ERIN, GREEN GEM OF THE SILVER SEA" (*U* 7.236) a specifically explicit link between "Aeolus" and "Cyclops" that in the end does come out in the published version. Although Joyce eventually drops his originally planned character list, I consider that he retains some "cross-exchange" between "Aeolus" and "Cyclops," and by the later added headline, along with the immediately following text, he reminds us of the connection between the two. The subhead "ERIN, GREEN GEM OF THE SILVER SEA" (*U* 7.236) is more than a rearrangement of recycled fragments; it brings out an overwhelming Revivalist voice that bombards, through Ned Lambert's reading from Dan Dawson's excessive speech in the newspaper, the listeners (readers) with a verbose nostalgic narrative on the landscape.

Vincent Cheng advises us that: "Dawson's rhetoric is in fact not unlike the linguistic excesses of the Celtic Revival which Joyce will parody in the 'Cyclops' episode"[48]. What Cheng calls the "Celtic Revival" is in fact, in Gibson's terms, synonymous with the Anglo-Irish Revivalist Movement, whose hypocritical pseudo-bardic style Joyce imitates and writes against in the Cyclopean passage of "Inisfail the fair." In correspondence to the pseudo-bardic style of Anglo-Irish pastorals, part of Dawson's speech reads:

> Or again, note the meanderings of some purling rill as it babbles on its way, tho' quarrelling with the stony obstacles, to the tumbling waters of Neptune's blue domain, 'mid mossy banks, fanned by gentlest zephyrs, played on by the glorious sunlight or 'neath the shadows cast o'er its pensive bosom by the overarching leafage of the giants of the forest. (*U* 7.243–47)

This fragment from Dawson's speech, bombastic and loquacious as it is, is a direct foreshadowing of, if not a parallel to, the Manganesque catalogue of "Inisfail the fair" in "Cyclops." Both passages consist of broad allusions to Ovid's *Metamorphoses* as well as references to the idyllic pastoral tradition. Gifford notes the lack of a specific, definitive source for this presumably fictional oration[49], and the fictitiousness of the piece makes it all the more significant, considering Joyce's massive dependency on contemporary subjects and journalism for his materials. The following section of this speech, as read out by Ned Lambert a few lines later, shifts the seemingly neutral (though verbose) praise of nature to attention on an Irish locale:

> As 'twere, in the peerless panorama of Ireland's portfolio, unmatched, despite their wellpraised prototypes in other vaunted prize regions, for very beauty, of bosky grove and undulating plain and luscious pastureland of vernal green, steeped in the transcendent glow of our mild mysterious Irish twilight ... That mantles the vista far and wide and wait till the glowing orb of the moon shine forth to irradiate her silver effulgence ... (U 7. 320–4, 327–8)

Such pompous praise of nature, and specifically Irish nature, will recur later in "Cyclops." In this way, Joyce's parody of the Anglo-Irish Revivalist pastoral in both passages creates a rhetorical connection between both episodes, foreshadowing and unleashing the tension between Anglo-Irish Revivalist rhetorics and the nationalistic one.

III.

Joyce's design of inter-exchange between the two passages from "Aeolus" and "Cyclops" does not simply rely on the rhetorical level. Gifford's note informs us: "Charles (Dan) Dawson was a successful baker who owned the Dublin Bread Company in Stephen's Street. He became one of Dublin's merchant-politicians: member of Parliament for County Carlow, lord mayor of Dublin (1882, 1883), and, *in 1904, collector of rates (taxes)* for the Dublin Corporation"[50]. Interestingly, the unnamed narrator of "Cyclops" is a debt collector, comparatively lower in society yet from the same line of business as a tax collector. Whereas the "Inisfail the fair" catalogue emerges following the unnamed debt collector's route toward Barney Kiernan's bar, the tax collector Dan Dawson's speech leads us into the excessive rhetoric of Anglo-Irish pastorals. Another, if less direct, link between the two passages resides in the profession of merchants and corresponding material consumption. Dan Dawson is a "merchant-politician" who starts his career as a bakery owner, and the character who reads his speech out of the newspaper in the *Evening Telegraph* office, Ned Lambert, coincidentally "works in a seed and grain store"[51], also in the line of commerce.

In "Cyclops," the unnamed narrator is a debt collector who works for Moses Herzog, a "grocer, or tea merchant"[52], another man of business.

It is worth noting that the Smithfield neighborhood surrounding St. Michan's parish, where the episode of "Cyclops" is set, has a history of vigorous commercial activities. Peter Pearson tells us that according to the Wide Streets Commissioners'[53] map archive, a significant amount of the houses located in this area had been used for trading since the 1800s. Not only was Church Street inhabited by merchants of diverse professions, the nearby Pill Lane was also "an important commercial district with similar trades," and among the residents were "a button manufacturer, a hemp and sail cloth merchant, a tobacco and snuff manufacturer, a linen draper, a hosier, pin-makers and a rope-maker trading under the name of Eliza Gibson"[54].

Where there is income, there is tax, and it is no exception in the late nineteenth century Dublin, whose economy relies heavily on trading. According to Brady and Simms, nineteenth-century Dublin functioned as a large regional capital with "industries and commerce, an important banking and insurance sector," with "elegant streets for shopping and enjoyment"[55]. However, the city's uneven distribution of population made it difficult to collect tax unhindered, especially in the metropolitan area. Brady and Simms point out that by then the "middle classes had moved to independent townships just beyond the borders of the 'city,'" and that the city, "as controlled by Dublin Corporation, was left with an unbalanced social and rateable structure," whereas "the great mass of the poor" who resided in the city "could contribute little to the coffers of the city"[56]. The few who had money, however, "lived beyond the reach of Dublin Corporation"[57]. The difficulty of rate collecting is also reflected in the notorious taxing actions undertaken in the general Smithfield area. Pearson writes:

> Up until the middle of the nineteenth century, it had been the role of the water bailiffs, who were employed by the city of Dublin, to collect customs or dues on 'horse, black cattle, sheep, lambs, calves and swine,' and also collected dues for the City on certain imports, such as wine and coal. ... However, in [1742] the City of Dublin decided that a collector would be appointed to collect the dues in Smithfield, in a system of subcontracting not unlike today's car clamping.[58]

Joyce's own speech in Trieste also reveals his knowledge of the menacing nature of tax collection as he informs the audience that "the soul of the country ... is paralysed by the influence and admonitions of the church, while its body is manacled by the police, the tax office, and the garrison" (*OCPW* 123). Charles (Dan) Dawson's speech, recited earlier in "Aeolus," is to mark his 1904 inauguration as the collector of taxes for the Dublin Corporation, whereas the Dublin

Corporation tax collecting system that is "not unlike today's car clamping" recalls the tax collector's counterpart, the nameless debt collector in "Cyclops," whom Frank Budgen terms a "snaring Thersites"[59].

In *Ulysses in Progress,* Groden indicates that Joyce "delayed that 'gigantic' passage [of 'In Inisfail the fair'] and introduced a new one before it"[60], referring to the contract between Moses Herzog and Michael E. Geraghty. In his view, "the question the passage raises [...] involves movement toward or away from the thing described (as the narrative moves closer to the document, the actual words become clearer)"[61]. The contract does function to let the readers move closer (toward the document) and away (from the document and toward the visible world). By such a process of closing in and backing up, readers are forced into and out of the compact space of the contract, and hence, the material transaction process under legal surveillance. The contract that outlines the financial dispute between Moses Herzog and Michael E. Geraghty is, on the one hand, a result of financial trade, and on the other, the product of power enforcement. It is not parodic in itself, but it paves the way for the subsequent Cyclopean catalogues to bombard, ridicule, or elaborate on the issue of imperial politics, and foreshadows a "social reading"[62] of the episode's setting as well as its cultural modernity.

When we read the contract as the incarnation of hierarchical control over monetary transactions, it becomes obvious that the contract is no longer simply an interrupting interpolation in the narrative. In fact, it comes out of the nameless narrator's stream of consciousness as his thoughts shift to his employer, Moses Herzog, and his coarse language when the debt collecting mission is assigned. In this view, and considering the tax collector's responsibility to collect rates in this area (as Pearson has mentioned), the narrator's voice that later appears between two gigantic catalogues is logical, though no less abrupt. Between the catalogue of vegetables and the catalogue of meat products, there emerges a line: *"I dare him,* says he, *and I double dare him. Come out here, Geraghty, you notorious bloody hill and dale robber!"* (*U* 12.100–1). These are the unnamed narrator's thoughts as his mind lingers around his debt collection task; it can also suggest the language of a menacing tax collector as he patrols around the market area.

These echoing coincidences help to explain and to contextualize Dawson's speech that in fact does not appear in the *Freeman's Journal* for 16 June 1904[63]. Joyce's parody of Anglo-Irish Revivalists' pseudo-bardic rhetorical style poses the question of a nationalistic narrative, whereas Joyce's insertion of the address, and especially when and where he locates this bombastic oration, adds layers to the interpretation of this passage. Dan Dawson, a merchant-politician who takes up the office of a tax collector in 1904, gives the fictional speech *"Our lovely land"*; Ned Lambert, a fictional employee of commerce, reads the script out of the newspaper in a sarcastic way. Since ridicule is very often politically assertive and hostile, what is implicit in the parallel is the confrontation of attitudes toward the politics of Irish nationalism. Such confrontation continues to develop, and occurs again in "Cyclops," between the unnamed narrator (the debt collector) and his encounter with the "bloody sweep" who almost drives "his gear into [the narrator's] eye" (*U* 12.2–3). The eye-blinding movement evokes the Homeric Cyclopean design for this episode and intensifies the narrator's counterpart position with the Citizen (another Cyclops), and yet, it furthermore warns the readers, at the very beginning of the entire episode, about the impending and contentious confrontation of nationalism and Irish politics that is to reach its climax later on.

IV.

The "ERIN, GREEN GEM OF THE SILVER SEA" (*U* 7.236) headline in "Aeolus" appears immediately after Bloom enters the *Freeman's Journal* office. Curiously, in "Cyclops" the mock-heroic catalogue starting with "Inisfail the fair" (*U* 12.68) also introduces the emergence of Bloom. In the V.A.8 copybook Joyce originally places a description of Bloom's movement between the "Inisfail" paragraph and "There rises a shining palace with crystal glittering roof, beheld from afar […]" (V.A.8, p. 1r):

> O'Bloom went [[illegible word]] Who comes through Inn's Quay ward, the parish of saint Michan. He moved a noble hero, [the son of Rudolph] [It is O'Bloom, the son of Rudolph, the son of Leopold Peter, son of Peter Rudolph he of the intrepid heart, impervious to all fear] eastward towards Pill lane, among the squatted [stench of] fishgirls and by the gutboards where lay heaps of red and purple fishguts. He went by the city market. O'Bloom [a man] of the intrepid heart [of gurnard, pollock, plaice, and halibut] (V.A.8, p. 1r)

The above introduction of Bloom's entrance will later be transferred into a more concise version, which is delayed until quite a few pages later (*U* 12.215–8), after the unnamed narrator's reflection on his earlier encounter with Bloom near the Dublin Vegetable and Fish Market.[64] His impression of Bloom "sloping around by Pill lane and Greek street with his cod's eye counting up all the guts of the fish" (*U* 12.213–4) in lieu of Joyce's initial straightforward description of Bloom's route "by the city market," "eastward towards Pill lane, among the squatted [stench of] fishgirls[65] and by the gutboards where lay heaps of red and purple fishguts" (V.A.8, p. 1r). This drastic change not only witnesses Joyce's stylistic transition in the middle stage during the writing process of *Ulysses* but also initiates a cataloguing style that, not necessarily well-classified, exhibits rather a commodifying system of listing.

Fritz Senn reminds us that "[o]riginally Joyce had started with fewer fish; the Rosenbach manuscript has only 'gunnard [sic], plaice, halibut, flounder' with 'the pollock' scribbled between the lines"[66]. He also notices the deviation in the "fishful streams": some of the names in the fish list are saltwater fish rather than freshwater ones. Reading the fish list rhetorically, Senn notes: "[a]fter a certain point the listing of fish as fish takes precedence over the original selective principle: 'streams' seems to have been lost sight of in favor of general fishfulness, which has developed its own generative momentum"[67]. Whereas Senn sees the act of enumeration in the fish list gradually obscuring the prerequisite setting of the "fishful streams" (*U* 12.71), I further regard the diversified fish species as a catalogue of fish on display in the Dublin Corporation Fruit, Vegetable and Fish Market. Phillip Herring, judging from the genetic materials, assumes Joyce copied "some of [the] lists of fish, vegetables, and other eatables" "from the market pages of Dublin newspapers"[68][69] and interpolated them into the Cyclopean catalogue. The fish list is consequently beyond the natural or the rhetorical; it becomes a catalogue of commodities for trade in the city market.

Located between St. Michan's parish and Arran Street, the Dublin Corporation Fruit, Vegetable, and Fish Market was opened in 1892, one of the propelling forces of its construction being the city's hygienic problems[70]. *The Irish Times* on 5 December 1892, the day before the market's opening, comments that the newly furnished construction would be "*splendidly* adapted" to accommodate the fish traders with its "immense stretch of ground, ... spanned by eight roofs, which supported on 56 cast iron columns and malleable arched iron girders". In the report presented to the Lord Mayor at the opening ceremony, the building is described as follows:

> ... The general style of the treatment is Romanesque. The elevation on Mary's Lane shows a central gateway flanked with the detached Corinthian columns of limestone, from Ballinasloe quarry, and on the other side are five arches of brick partly faced with terracotta. .... A dado of glazed brick, five feet high runs round the building, and the plain surface is relieved by bands of coloured bricks.

This is the Cyclopean "shining palace whose crystal glittering roof is seen by mariners who traverse the extensive sea in barks built expressly for that purpose" (*U* 12.87–89). Meadhbh Lysaght describes how, "[i]n the 1890s (and for many years afterwards), local farmers would bring their produce by horse and cart, delivering produce every day," and that "[g]oods would also be carted to the market from Dublin port (and from the railway stations in the case of the South

156   Pastoral

Fig. 4.1: A Panoramic of the Dublin Corporation Fruit, Vegetable, and Fruit Market. Image year unknown, The Irish Architectural Archive, Dublin.

Fig. 4.2: The Interior of the Dublin Corporation Market. Early 1980s, The Irish Architectural Archive, Dublin.

of Ireland produce ...)"[71]. So "thither come all herds and fatlings and first fruits of that land for O'Connell Fitzsimon[72] takes toll of them, a chieftain descended from chieftains" (*U* 12.89–91). Such are the activities engaged in the Dublin Corporation market at the time: transportation, commodity circulation, and institutional supervision.

The Dublin Corporation market is not unlike the "exhilarating site of modern consumption [...] represented by the electrically lit Araby bazaar"[73], which, as Leonard argues, appears as a site of desirable commodities for the boy consumer in "Araby." Unlike the consumerist desire elicited by the exotic Araby bazaar, in "Cyclops" the site of commerce -- the Dublin Corporation Wholesale Market, is rendered (though mockingly) heroic. I have mentioned earlier the Cyclopean fish catalogue's resemblance to the classified market news in Dublin newspapers, and such resemblance also applies to the following catalogues of vegetables, fruits, meat and dairy products in "Cyclops" (*U* 12.91–99, 12.102–117). The classified market news, at first glance, does not correspond to the emblematic code and the concise language usually expected in a piece of advertisement. However, bombarded with publicized lists of commodities, the market news conveys a commodious nature not unlike business advertisements. *Oxford English Dictionary* defines "advertisement" as a "public notice of announcement: *formerly* by the town-crier; *now,* usually, in writing or print, by placards, or in a journal; *spec.* a paid announcement in a newspaper or other print." Undoubtedly a "public notice of announcement," market news appropriately belongs to the category of advertisement, which, as Jennifer Wicke claims, is "a central means of cultural circulation and recirculation in modernity"[74]. In this sense, the Cyclopean catalogues of fish, fruits, vegetables, and other agricultural products have items accumulated for a reason: they display themselves as commodities in market circulation, in commercial display, and in cultural modernity, in a scale "too numerous to be enumerated" (*U* 12.74).

But is it true that Joyce may have in mind the concept of commercial activities and commodity circulations when he first devised, and later augmented, these catalogues of commercial produce? As I have pointed out via Groden's observations, the "Inisfail the fair" line was not in Joyce's initial paragraph when he first drafted the mock-heroic pastoral passage; it wasn't included in the paragraph until later, when Joyce substituted the original "green Erin of the west" with his Manganesque allusion. Although I hold on to my previous arguments on Joyce's perspective on Mangan and his "Prince Aldfrid's Itinerary through Ireland," the Manganesque addition, which comes later, cannot be Joyce's initial model for his first draft, let alone their discrepant structures. What, then, did he have in mind when he first composed the paragraph, which is significantly the first one he drafted for "Cyclops"?

An excerpt from Edmund Spenser's *A View of the State of Ireland*[75] strikingly resembles Joyce's mock-heroic list of fishfulness, trees, and the coming of faraway heroes and princes. Following a long recounting of devastating battles that ruined Ireland under the influence of Edward Bruce, brother of the King of Scotland, the passage shifts its focus to the natural beauty and natural resources of Ireland:

> And sure it is yet a most beautiful and sweet countrey as any is under heaven, being stored throughout with many goodly rivers, replenished with all sorts of fish most abundantly, sprinkled with many very sweet ilands and goodly lakes, like little inland seas, that will carry even shippers upon their waters, adorned with goodly woods even fit for building of houses and ships, so commodiously, as that if some Princes in the world had them, they would soone hope to be lords of all the seas, and ere long of all the world: also full of very good ports and havens opening upon England, as inviting us to come upon them, to see what excellent commodities that countrey can afford, besides the soyle it selfe most fertile, fit to yeeld all kinde of fruit that shall be committed thereunto. And lastly, the heavens most milde and temperate, though somewhat more moist then the parts toward the West.[76]

Apart from their mutually mirrored contents, their narrative structures also follow a similar order. Like the "fishful stream where sport ... and other denizens of the aqueous kingdom too numerous to be enumerated" (*U* 12.71, 74), Spenser observes in Ireland "many goodly rivers replenished with all sorts of fish most abundantly"[77]. As Joyce narrates "the lofty trees wave in different directions their firstclass foliage, the wafty sycamore, ... and other ornaments of the arboreal world with which that region is thoroughly well supplied" (*U* 12.75–78), the colonizer-speaker Irenaeus praises the country's abundance in wood supply with its "goodly woods fit for building of houses and ships so commodiously"[78].

In *A View*, Irenaeus recognizes Ireland's prospective commercial profit and land value as he depicts the country's "goodly woods even fit for building of houses and ships, so commodiously," and as that "if some Princes in the world had them, they would soone hope to be lords of all the seas, and ere long of all the world"[79]. This is not simply a parallel to Joyce's Inisfail catalogue with "heroes voyage from afar to woo them" (*U* 12.83). The heroes listed in Joyce's Inisfail paragraph include those "from Eblana to Slievemargy, the peerless princes of unfettered Munster and of Connacht the just and of smooth sleek Leinster and of Cruachan's land and of Armagh the splendid and of the noble district of Boyle, princes, the sons of kings" (*U* 12.83--86), place names and epithets directly borrowed from Mangan's "Prince Aldfrid's Itinerary through Ireland." In Spenser's version, the colonizer-speaker regards Ireland "full of very good ports and havens opening upon England," as if "inviting us to come upon them, to see what excellent commodities that countrey can afford, besides the soyle it selfe most fertile, fit to yeeld all kinde of fruit that shall be committed thereunto"[80]. Whereas Spenser's speaker reads the abundance of produce and potential commodity transactions with the colonial desire of an empire (in this case, of England), Joyce's list romanticizes merchants from all parts of Ireland who engage in commodity transactions in Dublin Corporation market.

Is it possible, then, that Joyce's plan for his princes of "Inisfail the fair" changes from Spenser's Princes of the empire to Mangan's princes of Ireland so as to evoke the question of colonial modernity as a response to anti-colonial voices and commodity culture of his time? Andrew Gibson describes the chapter of "Cyclops" as a "massive recycling of a stock-in-trade"[81], commercializing the Cyclopean lists. Jennifer Wicke, on the other hand, reads *Ulysses* as "the very best advertisement for modernism extant, *the colony's perfect revenge*," contending that "what it advertises is not the glory of British literature, but the velocity of decolonization brought about by the literary imagination, tunnelling from within the English language, and the very wide span a casual vision — or a social reading — can take in"[82]. By locating the colonial discourse of *A View* alongside the heroic rhetoric of "Prince Aldfrid's Itinerary through Ireland" in a setting of mass consumerist culture, Joyce seems to suggest the possibility of reading the Dublin Corporation market politically, (mass) culturally, and intertextually.

Joyce's rendering of the market allows us to interpret the marketplace as a colonial space — literally or commercially — that involves linear and horizontal confrontations of time and space. Enda Duffy's "Disappearing Dublin" interestingly appeals for attention to the "subaltern spaces" in Joyce's *Ulysses*. Taking from Lefebvre's[83] thesis of the capitalist abstraction of space, he points out that, according to Lefebvre:

> Capitalism relentlessly rationalizes space, that is, expunges its peculiarities and reminders of earlier forms of production to render it serviceable to exploitation, production, consumption, and the generation of profits. Capitalist controlled space tends to be abstracted space — which means spaces as bland as possible to facilitate the traffic in goods and workers, a space marked by prohibition, and, in ways Foucault delineated, pervasive surveillance.[84]

Duffy reminds us that apart from the dismissed specificities, certain places are also "erased" in *Ulysses*. He signals the prevalence of heterotopic[85] places (the cemetery on the city edge, the strand at Sandymount, the Ormond Hotel and Nighttown) as well as the absence of the "milieu of the colonist ruling cadres, whether in the Phoenix Park, Dublin Castle itself, or in the posher south Dublin suburbs, and the zones of the poorest workers, where Lefebvrian abstraction might be most in evidence"[86]. In his cutting-edge "Of Other Spaces," Michel Foucault brings up the idea of "heterotopia" and contemplates commodity accumulation through time and space:

> ... the idea of accumulating everything, of establishing a sort of general archive, the will to enclose in one place all times, all epochs, all forms, all tastes, the idea of constituting a place of all times that is itself outside of time and inaccessible to its ravages, the project

of organizing in this way a sort of perpetual and indefinite accumulation of time in an immobile place, this whole idea belongs to our modernity.[87]

Foucault's essay lists museums and libraries as examples of heterotopias, yet judging from the definition he provides, Joyce's Dublin Corporation market is undoubtedly among the heterotopias. In this view, and in light of Jennifer Wicke's contention, the Dublin Corporation Fruit, Vegetable, and Fish Market is both a space made commercial and an advertisement made spatial, a commodious space of accumulations and a colonial space of modernity.

David Spurr's "Colonial Spaces in Joyce's Dublin" proffers an exciting example of the "spaces of colonial authority"[88]. He explains that "in architectural parlance, a space of authority is one physically dominated by an imposing structure in such a way that it extends the area of that domination"[89], and in colonial Dublin, where imperial hierarchy dominates, he argues for the autocratic structure imposed upon architecture, rendering spaces colonially authoritative. The major focus of his article rests on the palace of the Four Courts, whose "menacing appearance" is "overwhelmed by its domineering architectural form, by its place in a history of colonial violence, and, in the context of *Dubliners*, by a more far-reaching repression internal to Irish society"[90]. To conclude, Spurr persuasively claims that: "For Joyce, British domination is part of both the architectural and the subjective environments; both architectural space and the space of consciousness are sites of a continual struggle among the competing claims of individual freedom, national aspirations, and imperial authority"[91]. Another construction under colonial supervision, the Dublin Corporation Fruit, Vegetable, and Fish Market likewise speaks of the imposed imperial authority over Dubliners. The *Irish Times* on the 5th of December, 1892, provides a detailed recount of the newly constructed exterior of the market:

> The front facing Halston street is a highly ornamental structure, of the Corinthian style, with columns, capitals, and entablature, these being surmounted by two artistically executed figures, the [first] representing Justice with the well-known sword, and the other "honest trade," holding in its hand a pair of evenly-balanced scales. The effect is heightened and localised by the presence of the City Arms in the centre — the three castles — this being accompanied by the motto, *"Obedientia civium urbis felicitas."*

The motto *"Obedientia civium urbis felicitas"* (Happy the city whose citizens obey) conveys an imperial demand for civil obedience, while the article's language discloses further authoritarian governance as "the effect is *heightened* and *localised* by the presence of the City Arms in the centre — the three castles — this being accompanied by the motto"[92]. The press release on the market's opening ceremony discloses further traces of institutional authority, including the Lord Mayor's praise of the market building as "second to none in the Empire." It is also worth nothing that, during the ceremony, a set of silver keys with shamrock embellishment was presented to the Right Hon. Lord Mayor of Dublin (see Fig. 4.3). Whereas the keys may remind us of Bloom's composition of the Keyes advertisement, on occasion such as the inauguration of Dublin Corporation Market, the symbolic keys submitted to the Lord Mayor illustrate a declared yielding to authoritarian governance. With the menacing rate collecting of market transactions, the oppressive governmental demand for obedience, the submission of the symbolic "keys" (hence the power to enter and to rule) to the authorities concerned (by then an Englishman or a newly risen Anglo-Irish middle-class), the Dublin Corporation market, initiated with the aim to solve Dublin's hygienic problems, consequently becomes a colonial space in which authoritarian and imperial controls predominate.

In Joyce's critique of the English Liberal Party's irresponsible treatment of the Home Rule Bill in his 1907 essay "Home Rule Comes of Age," he signals that the Home Rule measure, even if carried out, would nonetheless give the Dublin executive council "no legislative power, no power to fix or control taxes, no control over thirty-nine of the forty-seven government offices, including those of the constabulary and the police, the supreme court or *the agrarian commission*" (*OCPW* 143, my italics). In other words, in colonial Dublin at the turn of the century, Joyce regards the agrarian commission along with the supreme court among the dominating vehicles of the British governmental rule. Joyce's concern, I argue, is reflected in his political choice in beginning the episode of "Cyclops" with a bullying rate collector not unlike English tax control, as well as a series of mock-heroic commercial market catalogues alluding to the city market in which the activities of agrarian commissions take place. In this view, the Dublin Corporation Fruit, Vegetable and Fish Market is visibly a colonial space not simply due to the colonial background during its construction and operation, but more so indicative of its colonial spatiality as agrarian commissions, Spenserian parody, and Manganesque allusion are all compressed in the mock-bardic catalogue of "Inisfail the fair."

V.

From Romantic pastoral tradition, Manganesque allusion, mock-bardic Anglo-Irish Revivalist writings, social languages, the language of advertisement, to Spenserian colonial discourse, the catalogue of "Inisfail the fair" orchestrates a passage of "social heteroglossia"[93] (Bakhtin) in which "dialogical contacts" between its diverse stratifications of languages occur. Bakhtin suggests how

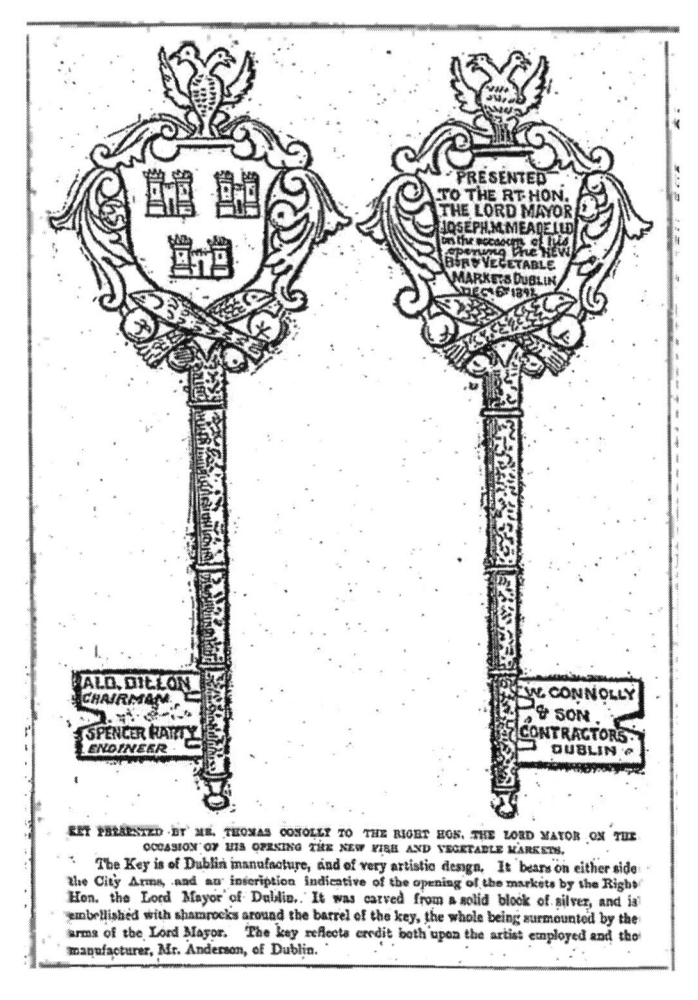

Fig. 4.3: A reprinted copy of the keys presented to the Lord Mayor at the opening ceremony of the Dublin Corporation Market. The description below the image reads: "KEY PRESENTED BY MR. THOMAS CONOLLY TO THE RIGHT HON. THE LORD MAYOR ON THE OCCASION OF HIS OPENING THE NEW FISH AND VEGETABLE MARKETS. The Key is of Dublin manufacture, and of very artistic design. It bears on either side the City Arms, and an inscription indicative of the opening of the markets by the Right Hon. the Lord Mayor of Dublin. It was carved from a solid block of silver, and is embellished with shamrocks around the barrel of the key, the while being surmounted by the arms of the Lord Mayor. The key reflects credit both upon the artist employed and the manufacturer, Mr. Anderson, of Dublin." (*Weekly Irish Times* 1892: 5)

"the languages that are crossed in it relate to each other as do rejoinders in a dialogue; there is an argument between languages, an argument between styles of language. But it is not a dialogue in the narrative sense, nor in the abstract sense; rather it is a dialogue between points of view, each with its own concrete language that cannot be translated into the other"[94]. Although, as Keith Booker says, "we certainly do not need Bakhtin's theories to recognize that Joyce draws extensively upon the entire western literary tradition"[95], Bakhtin's theories assist us in reading Joyce's cross-referential languages as interactive agents of dialogic interchanges. At the beginning of "Cyclops," with the series of catalogues starting with "Inisfail the fair" (*U* 12.68–117), Joyce provides the readers with an exemplary dialogic mosaic of language stratifications and social heteroglossia. Whereas these — literary, historical, social, political, and even commercial — discourses converge in a complex way that brings out the multiple voices and ideologies of the dialogue, the passage's multiple layers of narrative provide possible readings of the catalogue that are by no means exclusively unitary.

In Michael J. McDowell's essay on Bakhtin and ecocriticism, he notes how "Bakhtin identifies the idyll as a model for restoring 'folkloric time'"[96]. Joyce's use of the idyll, as this chapter has attempted to demonstrate, serves far more than simply to "[restore] folkloric time." By responding to an Anglo-Irish pastoral tradition of nature writing, Joyce's idyllic illustrations of nature are located intricately in contexts that trigger questions on the politics of the idyllic language against the background of Anglo-Irish Revival. On the other hand, they also bring out the dynamics of the languages of capitalism, and modern transactions in the marketplace as well as among multiple layers of social languages. As Bakhtin points out, in writings of the idyll, "nature itself [has] ceased to be a living participant in the events of life, [but has been] fragmented into metaphors and comparisons serving to subliminate individual and private affairs and adventures not connected in any real or intrinsic way with nature itself"[97]. Whereas pastoral writings of the landscape and the idyll seem to bring readers to the escapist surroundings of nature, they in fact, as in the case of *Ulysses*, awaken diverse levels of critical considerations for the modern relationship between languages and eco-politics.

Notes

1. Emer Nolan, *James Joyce and Nationalism* (London: Routledge, 1995), p. 106.
2. According to Joyce's schema, the "art" for this chapter is "politics" and the "symbol" "Fenian," terms which not only relate to politics, but involve Irish

independent movements associated with Irish nationalism (see Stuart Gilbert, *James Joyce's "Ulysses"*, New York, 1930; revised, 1952).
3. M. M. Bakhtin, *The Dialogic Imagination: Four Essays*, trans. by Vadim Liapunov and Kenneth Brostrom, eds. by Michael Holquist and Vadim Liapunov (Austin: University of Texas Press, 1982), p. 312.
4. Christopher Marlowe, "The Passionate Shepherd to his Love."
5. Schiller, "On Naïve and Sentimental Poetry," p. 210.
6. Roger Sales, *English Literature in History 1780-1830: Pastoral and Politics* (London: Hutchinson, 1983), p. 17.
7. Raymond Williams, *The Country and the City* (Oxford University Press, 1973), p. 60.
8. Greg Garrard, *Ecocriticism* (New York: Routledge, 2004), p. 35.
9. Gifford, p. 316.
10. Mangan, p. 194.
11. Ibid., pp. 194–96.
12. See Book IX. of *The Odyssey of Homer: Done into English Verse* by William Morris. However, similar passages on the natural abundances of the Cyclops' island can also be found in different versions of *The Odyssey* in Book IX.
13. Brigitte Sandquist's contention on Ovidian metamorphoses in her paper "The Tree Wedding in 'Cyclops' and the Ramification of Cata-logic" may also be placed among the interwoven narrative schemes I here propose.
14. Terry Eagleton, *Heathcliff and the Great Hunger: Studies in Irish Culture* (Verso, 1995), p. 274.
15. Ibid.
16. For more on unevenly developed post-Famine Ireland, see Terry Eagleton's *Heathcliff and the Great Hunger*, "The Archaic Avant-Garde" (Ch.2), or Cormac O Grada's *Ireland Before and After the Famine*, p. 27, 29.
17. Eagleton, p. 7.
18. Sean Ryder, "Introduction," in *James Clarence Mangan: Selected Writings* (Dublin: University College Dublin Press, 2004), pp. 9–10.
19. Ibid., p. 9.
20. Ibid., p. 5.
21. Taken and elaborated in his *James Clarence Mangan: Selected Writings* introduction by Sean Ryder from *Dublin University Magazine* (April 1839), 494. (Ryder 6).
22. Ryder, p. 8.
23. Kilroy, p. 45.
24. Ryder, p. 467.
25. Joyce illustrates how Mangan, "[a]lthough he wrote such admirable English, [...] refused to work for English magazines or journals" (*CW* 134). By doing so Mangan negatively rebels against the British empire by refusing to have his writings published by English publishers.

26. Marjorie Howes, "'Goodbye Ireland I'm going to Gort': Geography, Scale, and Narrating the Nation," in *Semicolonial Joyce*, eds. by Derek Attridge and Marjorie Howes (Cambridge: Cambridge University Press, 2000), p. 59.
27. Ibid.
28. Michael Groden, *Ulysses in Progress* (Princeton: Princeton University Press, 1977), p. 115.
29. Ibid., p. 124.
30. This copybook, numbered MSS 36,639/10 by the National Library of Ireland, appeared and was purchased by NLI in 2002. See Groden's "The National Library of Ireland's New Joyce Manuscripts," *Joyce in Trieste: An Album of Risky Readings*, for introductory descriptions of the manuscript (as well as other materials simultaneously obtained). For specific information on the NLI "Cyclops" collection, see Groden's "Joyce at Work on 'Cyclops': Toward a Biography of *Ulysses*," *JJQ* 44.2 (2007): 217–45.
31. Sylvia Beachconsiders V.A.8 copybook as the earliest version of "Cyclops," and Groden's *Ulysses in Progress,* as well as his survey on the new NLI collection, confirms that.
32. Phillip F. Herring, ed., *Joyce's Notes and Early Drafts for* Ulysses: *Selections from the Buffalo Collection* (Charlottesville: University of Virginia Press, 1977), p. 132.
33. Gibson, *Joyce's Revenge*, p. 107.
34. Ibid., pp. 107, 118.
35. Ibid., p. 118.
36. Lists in "Cyclops," according to Groden, were added afterwards and continued to expand until the publication of *Ulysses* in 1922 (1977: 164–5).
37. Thornton's suggestion on the influence of Mangan's poem "Prince Aldfrid's Itineray Through Ireland" (*Allusions in "Ulysses": an Annotated List*, UNC Press, 1982, p. 256-7) corresponds with Gifford's note in *Ulysses Annotated: Notes for James Joyce's* Ulysses (University of California Press, 1974), p. 316.
38. Herring, pp. 132-3.
39. William M. Schutte, "An Index of Recurrent Elements in 'Ulysses': 'Cyclops,'" in *JJQ*, Vol. 16, No. 1/2, Structuralist/Reader Response Issue (Fall, 1978 – Winter, 1979), p. 164.
40. Gibson, p. 125.
41. Ibid.
42. See Groden, "'Cyclops' in Progress, 1919," in *JJQ*, Vol. 12, No. 1/2, Textual Studies Issue (Fall 1974 – Winter 1975), p. 127.
43. Groden, *Ulysses in Progress*, p. 107.
44. In parallel, rather than in conflict, with my argument are the possible literary sources for this headline: both Gifford's notes and, earlier, Mabel P. Worthington's *PMLA* article, mention the association with Thomas Moore's

"Let Erin Remember the Days of Old" ("Ere the emerald gem of the western world") and the Shakespearean description of England in *Richard II* II.i.: "This precious stone set in a silver sea." (Worthington 325; Gifford 133)
45. Groden, *Ulysses in Progress,* p. 112.
46. Ibid., p. 133.
47. Ibid.
48. Cheng, p. 186.
49. Gifford, p. 133.
50. Ibid., p. 107, my italics.
51. Gifford, p. 106.
52. Ibid., p. 314.
53. "The Commissioners for making Wide and Convenient Ways, Streets and Passages in the City of Dublin." Established by an Act of Parliament in 1757, the Commission "had extensive powers with the authority to acquire property by the compulsory purchase, demolish it, lay down new streets and set lots along the new streets to builders for development." It was abolished in 1851, under the Dublin Improvement Act of 1849. The commissioner's collection includes 800 manuscripts maps that document Dublin before, during, and after the work of the commission. (quotations and information from Dublin City Council website:<http://www.dublincity.ie/>)
54. Peter Pearson, *The Heart of Dublin: Resurgence of an Historic City* (Dublin: The O'Brien Press, 2000), p. 379.
55. Joseph Brady and Anngret Simms, "Dublin in the Nineteenth Century: an Introduction," in *Dublin Through Space and Time (c.900-1900),* eds. By Brady and Simms (Dublin: Four Courts, 2011), p. 160.
56. Ibid., pp. 160–61.
57. Ibid., p. 161.
58. Pearson, pp. 385–6.
59. Frank Budgen, *James Joyce and the Making of "Ulysses"* (Bloomington: Indiana University Press, 1960), p. 154.
60. Groden, *Ulysses in Progress,* p. 154.
61. Ibid.
62. The term is framed by Jennifer Wicke in her book *Advertising Fictions: Literature, Advertisement, and Social Reading* (Columbia University Press, 1988).
63. Gifford, p. 107.
64. See Gifford's note 12.87 (294.18), p. 318, as well as note 12.213-4 (297:36–37), p. 326.
65. This quotation is taken from Myron Schwartzman's transcription of the V.A.8 Copybook in "The V.A.8 Copybook: An Early Draft of the 'Cyclops' Chapter of '*Ulysses*' with Notes on its Development" (*JJQ* 12:1/2, 64–122). Yet Schwartzman's transcription differs from Herring's (in *Joyce's Notes and*

Early Drafts for Ulysses: *Selections from the Buffalo Collection*) in the line "among the squatted [stench of] fishgirls" (V.A.8 p. 1r, Schwartzman 71). Instead of "fishgirls," Herring's transcription has the "squatted {stench of} *fishgills*" (V.A.8 p. 1r, Herring 135, my italics). Such discrepancy may be due to the illegibility of Joyce's handwriting. However, despite his transcription of "fishgills" instead of "fishgirls," Herring also recognises the presence of fishwives in this early draft. He comments how: "First Joyce painted a highly idealistic portrait of the Dublin Fish and Vegetable Market *with its attendant fishwives and customers*, then parted the clouds briefly to reveal 'squatted {stench of} fishgills' and the 'heaps of red and purple fishguts' and to allow us a glimpse into the mind of Leopold Bloom, a sign that Joyce once considered using the interior monologue style of other episodes" (Herring 132-3, my italics).

66. Fritz Senn, "Ovidian Roots of Gigantism in Joyce's 'Ulysses'" in *Journal of Modern Literature* 14.4 (Spring 1989), p. 563.
67. Ibid.
68. Herring, p. 133.
69. A sample of the "fish" section on classified market news of *Freeman's Journal* in late 19th century looks like this: "DUBLIN (WHOLESALE) -- May 13 -- There was a big supply of fish on market this morning. Arrivals included a fair catch of local trawl fish, large hooked fair country deliveries, and about 53 boxes of stuff from a couple of steamers. Salmon, 1s 9d to 2s 2d per lb; best lobsters, 12s 6d to 15s and 18s per13; others, 3s to 8s and 10s per lot; mackerel, 15s to 16s and 17s per 1/2 box; turbet, 4s to 10s and 12s 6d each; brill, 1s to 3s each; soles, 20s to 30s, and 32s per 9 pair; medium, 10s to 17s 6d; small, 4s to 8s per 9 pair; dory, 6d to 1s each; large plaice, 4s to8s per 8; others, 1s to 5s per doz; prime hooked cod, 8s to 12s per 4; average run, 2s to 6s per 4; haddock, 2s to 3s per 4; a few special selected, 4s and 6s per 4; average run, 2s to 6s per 4; small, 8d to 2s per 8 and 12; hake, 1s to 2s6d each; ling, 6d to 1s 6d each; conger eels, 1s to 2s 6d each; white soles, 1s to 1s 6d per 6; whiting, 1s to 2s 6d per bank; red gurnard, 6d to 1s each; rough fish went cheap." (*Freeman's Journal and Daily Commercial Advertiser,* May 14, 1898)
70. In an *Irish Times* article that releases the news of the forthcoming city market, the market is anticipated as "the vast benefit in a sanitary sense, which will be conferred on a thickly populated district by the clearing away of sites which were synonymous with filth, while the old markets never at any time fulfilled the requirements of public marts for the purchase and sale of necessary articles of food" (*Irish Times* 1889: 3; *Weekly Irish Times* 1889: 6).
71. Meadhbh Lysaght, "Trinity College Schools' Competition Junior Gold Medal Winner: Dublin's Wholesale Fruit & Vegetable Market" in *History Ireland* 4:3 (Autumn 1993), pp. 43-44.

72. According to Gifford's note, O'Connell Fitzsmion served as the superintendent of the food market in 1904, and is responsible for the tolls collected from the market. (Gifford 318)
73. Garry Leonard, "Joyce and Advertising: Advertising and Commodity Culture in Joyce's Fiction" in *James Joyce Quarterly*, Vol. 30/31(Summer – Fall, 1993), p. 575.
74. Jennifer Wicke, "Modernity Must Advertise: Aura, Desire, and Decolonization in Joyce" in *James Joyce Quarterly* 30:4 - 31:1 (Summer – Fall 1993), p. 609.
75. This controversial anti-Irish book is a lengthy dialogue between Eudoxus, "a rational Englishman interested in politics but largely ignorant of Ireland," and Irenaeus, who is "clearly speaking from the position of knowledge and probably represents one of the New English colonists, like Spenser himself" (Hadfield xvii).
76. Edmund Spenser, *A View of the State of Ireland: from the first printed edition* (1633), eds. by Andrew Hadfielf and Willy Maley (Oxford: Blackwell Publishers, 1997), p. 27.
77. Ibid.
78. Ibid.
79. Ibid.
80. Ibid.
81. Gibson, *Joyce's Revenge*, p. 117.
82. Jennifer Wicke, "Joyce and Consumer Culture" in *The Cambridge Companion to James Joyce* (Cambridge University Press, 2004), p. 241, my italics.
83. Lefebvre, *The Production of Space*, trans. by Donald Nicholson-Smith (Oxford: Basil Blackwell, 1991).
84. Enda Duffy, "Disappearing Dublin: *Ulysses*, Postcoloniality, and the Politics of Space," in *Semicolonial Joyce* (Cambridge University Press, 2000), pp. 42–3.
85. The term "heterotopia" is introduced by Michel Foucault in "Of Other Spaces," in which he defines it as "places that do exist and that are formed in the very founding of society — which are something like counter-sites, a kind of effectively enacted utopia in which the real sites, all the other real sites that can be found within the culture, are simultaneously represented, contested, and inverted. Places of this are outside of all places, even though it may be possible to indicate their location in reality" (Foucault 24)
86. Duffy, p. 53.
87. Michel Foucault and Jay Miskowiec, "Of Other Spaces," in Diacritics 16:1 (Spring 1986), p. 26.
88. David Spurr, "Colonial Spaces in Joyce's Dublin," in *James Joyce Quarterly* 37:1/2, Dublin and Dubliners (Fall 1999 – Winter 2000), p. 24.
89. Ibid.
90. Ibid, p. 26.
91. Ibid., p. 24.

92. Italics mine.
93. Whereas there is internal stratification "in every language at any given moment of its historical existence" (Bakhtin 263), heteroglossia is "the base condition governing the operation of meaning in any utterance" (Bakhtin 428) in each internal stratified utterance. Heteroglossia enables a "multiplicity of social voices and a wide variety of their links and interrelationships between utterances and languages" (Bakhtin 263). On the other hand, dialogism is "the characteristic epistemological mode of a world dominated by heteroglossia" (Bakhtin 426).
94. Bakhtin, p. 76.
95. M. Keith Booker, *Joyce, Bakhtin, and the Literary Tradition: Toward a Comparative Cultural Poetics* (Ann Arbor: University of Michigan Press, 1997), p. 10.
96. Michael J. McDowell, "The Bakhtinian Road to Ecological Insight," *The Ecocriticism Reader: Landmarks in Literary Ecology*, eds. by Cheryll Glotfelty and Harold Fromm (Athens: University of Georgia Press, 1996), p. 379.
97. Bakhtin, qtd. by McDowell, pp. 378–9.

CONCLUSION
The Journey and Beyond

The beginning of this project took form back in 2007, when I was enrolled in Professor Michael Levenson's postgraduate module on "*Ulysses* and Modernism" during my studies in the University of Virginia for my Master's degree. The module took us through all the chapters of *Ulysses* alongside other texts and theories of modernism, and through such parallel readings we spent more than a dozen weeks discussing different modernist expressions of art and languages. What caught my attention in particular were the expressions of the environment, such as trees, landscape, herds, gardens, garbage, etc., in these modernist writings, especially in *Ulysses*. As a gesture toward solving my inquiries, by the end of the term I wrote a twenty-page paper to address the question of the environment in *Ulysses*. Unbeknown to me, instead of answering the questions I had in mind, the research I conducted for that particular essay introduced even more questions which I couldn't find satisfactory answers for at the time. Thus began my journey of taking on the challenge of delving deeper into understanding the (eco)politics of Joyce's writing of nature and environment in *Ulysses*.

This study is the result of that prolonged journey in pursuit of understanding the concepts of the garden, sewage, trees, landscape, and the marketplace, in the Joycean context. Initiated at the time when ecocriticism was still considered mainly irrelevant to modernism or Joyce studies, let alone *Ulysses* with its urban setting, this project has attempted to examine elements of nature in *Ulysses* through close readings and reconsiderations of nature, and hence ventured the possibility of an eco-political reading via the lenses of history, politics, languages and cultural politics. Among others, the archival resources from recently flourishing Joycean genetic criticism have significantly contributed to my reading and interpretation of Joyce's writing on nature and specific ecological topics in *Ulysses*. They have helped me locate the environmental impact on Joyce during his compositional process. Having made use of resources from genetic studies on Joyce's manuscripts for all of my four chapters, I was able to discover further evidence of Joyce's awareness of contemporary environmental issues (such as sewage problems or afforestation propaganda), as well as his intentional use of natural/idyllic languages to reveal, parody, and satirise modern Irish questions of nationalism and consumerism. By specifically focusing on the cultural politics of the Anglo-Irish Co-operative recommendation of an ideal country life, the traumatic Irish history of land distribution, and the rise of the culture of consumption in Ireland at the turn of the century, the first part of this project has attempted to question the accepted images of nature in urban fin-de-siècle Dublin. Moreover, the ideal noble farmer as propagandized in *The Irish Homestead*, the utopian picture of the garden city suburbs, and the leaflet of a Palestinian utopia have suggested an alternative reading of material waste not simply as cultural artifacts or disposables of social infrastructure but as an alternative narrative voices of the city. The latter part of the book moves to the questions of literary representations of nature and social languages, and, through close readings of the tree wedding passage in "Cyclops" and in the pastoral representation of landscapes, has proposed a reconsideration of the ecosystem in light of languages and cultural politics in relation to nationalism and nature. This study is by no means a comprehensive eco-political study of *Ulysses*. Rather, what it hopes to do is to suggest the possibility of reading the environment alongside history, politics, social languages, and genetic criticism, and by doing so to locate the interpretation of eco-politics in a Joycean context that is historically and socio-politically relevant.

To return to an earlier quotation by Robert Pogue Harrison: "History without gardens would be a wasteland. A garden severed from history would be superfluous"[1]. This study began with a series of questions to examine more deeply the material notions of gardens, sewage, trees, and landscape as stylistic representations of nature, and along the journey, the interwoven dynamic among history, politics, and nature started to unfold. One of the most frequent questions I have been asked regarding my research is that, since I am proposing an ecocritical reading of his work, whether I would consider James Joyce an ecological writer. Having looked into several thematic environmental issues addressed in *Ulysses* with the evidence of historical and archival resources over the course of this project, I believe this book has demonstrated that Joyce, though not necessarily an "ecological writer," is after all a writer with the environment in mind, and that the imagination of nature in *Ulysses* is in fact inseparable from that of the emergent nation of fin-de-siècle Ireland.

What does an eco-political reading of *Ulysses* look like? If Joyce's own ecosystem of languages of nature, nation, and consumption, is where we find "[t]he war […] in words and the wood [as] the world" (*FW* 98.34–5), then it is probably fair to say Joyce has redeemed the superfluousness of his then-contemporary Anglo-Irish Revivalist nature writing through his art of composing nature in the most unlikely setting of "dear dirty Dublin".

Note

1. Harrison, p. x.

List of Figures

Fig. 1.1:	Foremen's cottages at "Kilkenny Garden Village" at Sheestown, Co. Kilkenny (1907), designed by William A. Scott. The illustration originally appeared in the prestigious Irish architectural periodical, *Irish Builder and Engineer*, on November 30, 1907. Source: archive.org/details/irishbuilderengi4919unse	47
Fig. 1.2:	**The Town-country Magnets.** Howard's idea of the three town-country magnets, illustrating what he perceives as the pros and cons of town and country livings, and ultimately, the advantages of a town-country, namely his garden city model. Source: Ebenezer Howard, *Garden Cities of To-morrow*	52
Fig. 2.1:	Adopted from "The Jewish Colonies in Palestine" from *Zionism and the Jewish Future*, pp. 138–9	76
Fig. 3.1:	Crowds of Spectators Observing a Parade Passing Through O'Connell Street. Picture taken by Clarke, J. J. Published between 1897–1904. Courtesy of the National Library of Ireland. <http://catalogue.nli.ie/Record/vtls000168779> Accessed 16 June 2012	115
Fig. 3.2:	INF brass band in Monaghan, 1910. Picture courtesy of Trinity College Library Dublin	116
Fig. 3.3:	INF banners with figures of Erin. Courtesy of Trinity College Library Dublin	118
Fig. 3.4:	INF banners with figures of Erin. Courtesy of Trinity College Library Dublin	119
Fig. 3.5:	INF members holding a banner of theirs, Ballyholland, August 1995. Seen in Jarman, *Material Conflicts*, p. 30	120
Fig. 4.1:	A Panoramic of the Dublin Corporation Fruit, Vegetable, and Fruit Market. Image year unknown, The Irish Architectural Archive, Dublin	156
Fig. 4.2:	The Interior of the Dublin Corporation Market. Early 1980s, The Irish Architectural Archive, Dublin	156

Fig. 4.3:   A reprinted copy of the keys presented to the Lord Mayor at the opening ceremony of the Dublin Corporation Market. The description below the image reads: "KEY PRESENTED BY MR. THOMAS CONOLLY TO THE RIGHT HON. THE LORD MAYOR ON THE OCCASION OF HIS OPENING THE NEW FISH AND VEGETABLE MARKETS. The Key is of Dublin manufacture, and of very artistic design. It bears on either side the City Arms, and an inscription indicative of the opening of the markets by the Right Hon. the Lord Mayor of Dublin. It was carved from a solid block of silver, and is embellished with shamrocks around the barrel of the key, the while being surmounted by the arms of the Lord Mayor. The key reflects credit both upon the artist employed and the manufacturer, Mr. Anderson, of Dublin." (*Weekly Irish Times* 1892: 5) 162

Bibliography

Primary Texts

Joyce, James. *Finnegans Wake*. New ed. New York: Penguin, 2015.

Joyce, James. *Occasional, Critical, and Political Writing*. Oxford University Press, 2008.

Joyce, James. *Ulysses*. Ed. Hans Walter Gabler. New York: Vintage, 1993.

Joyce, James. *Selected Letters of James Joyce*. Ed. Richard Ellmann. New York: Viking, 1975.

Archival Resources

Dublin Corporation. *Report of the Markets Construction Committee*, 6 December 1892.

Great Britain. Department of Agriculture and Technical Instructions for Ireland. *Report of the Departmental Committee on Irish Forestry*. Dublin: HMSO, 1908. *House of Commons Parliamentary Papers (HCPP)*. Web. 31 Oct. 2010. <https://www.proquest.com/products-services/House-of-Commons-Parliamentary-Papers.html>

"Guide to Pre 1840 Collections III: Wide Streets Commissioners 1758 - 1851." Dublin City Council, n.d. Web. <http://www.dublincity.ie/> 20 March 2010.

Herring, Phillip F., ed. *Joyce's Notes and Early Drafts for* Ulysses: *Selections from the Buffalo Collection*. Charlottesville: University of Virginia Press, 1977.

Herring, Phillip F., ed. *Joyce's* Ulysses *Notesheets in the British Museum*. Charlottesville: University of Virginia, 1972.

Joyce, James. *The Finnegans Wake Notebooks at Buffalo VI.B.3, VI.B.6, VI.B.10, VI.B.14, VI.B.25, VI.B.29*. Vincent Deane, Daniel Ferrer, and Geert Lernout, eds. Turnhout: Brepols, 2001–2002.

"Market News." *Freeman's Journal and Daily Commercial Advertiser* (Dublin, Ireland), Saturday, 14 May 1898. *Irish Newspaper Archives*. Web. 13 March 2010. <https://www.irishnewsarchive.com/>

"New Fish and Vegetable Market." *Irish Times* [Dublin] 7 Dec. 1892: 6. *Irish Times*. Web. 13 March 2010. <https://www.irishtimes.com/archive>

"Pill Lane, the New Fish Market, and the New Street." *Irish Times* [Dublin] 28 Sep. 1894, 8. *Irish Times*. Web. 13 March 2010. <https://www.irishtimes.com/archive>

"The New Vegetable and Fish Markets." *Irish Times* [Dublin] 12 June 1889: 3. *Irish Times*. Web. 13 March 2010. <https://www.irishtimes.com/archive>

"The New Vegetable and Fish Markets." *Weekly Irish Times* [Dublin] 15 June 1889: 6. *Irish Times*. Web. 13 March 2010. <https://www.irishtimes.com/archive>

"The New Fish and Vegetable Markets." *Irish Times* [Dublin] 5 Dec. 1892: 5. *Irish Times*. Web. 13 March 2010. <https://www.irishtimes.com/archive>

"The New Fish and Vegetable Markets." *Weekly Irish Times* [Dublin] 10 Dec. 1892: 5. *Irish Times*. Web. 13 March 2010. <https://www.irishtimes.com/archive>

Secondary Works

Abercrombie, Patrick, Sydney Kelly, and Arthur Kelly. *Dublin of the Future: the New Town Plan*. Liverpool: University Press of Liverpool, 1922.

Albion, R. G. *Forests and Sea Power: The Timber Problems of the Royal Navy 1562-1826*. Cambridge: Harvard University Press, 1926.

Allen, Nicholas. *George Russell (AE) and the New Ireland*. Dublin: Four Courts, 2003.

Alloway, Lawrence. "Junk Culture." *Architectural Design* 31/3 (March 1961): 123-33.

Arnold, Matthew. "On the Study of Celtic Literature." *Lectures and Essays in Criticism*, ed. R. H. Super. Ann Arbor: University of Michigan Press, 1962. Pp. 291-395.

Bakhtin, M. M. *The Dialogic Imagination: Four Essays*. Trans. by Vadim Liapunov and Kenneth Brostrom. Eds. by Michael Holquist and Vadim Liapunov. Austin: University of Texas Press, 1982.

"Bark." Entry 1. *Oxford English Dictionary*. 2nd ed. 1989. Web. Accessed 01 March 2010. <www.oed.com>

Baron, Naomi S., and Nikhil Bhattacharya. "Vico and Joyce: the Limits of Language." *Vico and Joyce*, Donald Phillip Verene, ed. Buffalo: SUNY Press, 1987. 175-95.

Beckett, Samuel. "Dante... Bruno. Vico... Joyce." *Our Exagmination Round His Factification for Incamination of Work in Progress*. New York: New Directions, 1962.

Bell, M. David. "The Search for Agendath Netaim: Some Progress, but No Solution." *JJQ* 12.3 (Spring 1975): 251-258.

Bender, Abby. *Israelites in Erin: Exodus, Revolution, and the Irish Revival.* Syracuse: Syracuse University Press, 2015.

Benstock, Bernard. "Vico... Joyce. Triv.. Quad." *Vico and Joyce*, Donald Phillip Verene, ed. Buffalo: SUNY Press, 1987. 59–67.

Black, Martha Fodaski. "Joyce on Location: Place Names in Joyce's Fiction." *Joyce and the City: the Significance of Place.* Ed. by Michael H. Begnal. Syracuse University Press, 2002. 18–34.

Bookchin, Murray. *The Limits of the City.* 2nd rev. ed. Montreal: Black Rose Books, 1986.

Bookchin, Murray. *Post-Scarcity Anarchism.* Montreal: Black Rose Books, 1986.

Booker, M. Keith. *Joyce, Bakhtin, and the Literary Tradition: Toward a Comparative Cultural Poetics.* Ann Arbor: University of Michigan Press, 1997.

Bosinelli, Rosa Maria. "'I use his cycles as a trellis': Joyce's Treatment of Vico in *Finnegans Wake*." *Vico and Joyce*, Donald Phillip Verene, ed. Buffalo: SUNY Press, 1987. 123–31.

Brady, Joseph, and Anngret Simms. "Dublin in the Nineteenth Century: an Introduction." *Dublin Through Space and Time (c.900-1900).* Eds. by Brady and Simms. Dublin: Four Courts, 2011.

Brazeau, Robert and Derek Gladwin, eds. *Eco-Joyce: the Environmental Imagination of James Joyce.* Cork: Cork University Press, 2014.

Brodetsky, Dr. S. "Cultural Work in Palestine." *Zionism and the Jewish Future.* Ed. by H. Sacher. London: John Murray, 1917. 171–189.

Buckley, Anthony D. "'On the club': Friendly Societies in Ireland." *Irish Economic and Social History* 14 (1987): 39–58.

Budgen, Frank. *James Joyce and the Making of "Ulysses."* Bloomington: Indiana University Press, 1960.

Buell, Lawrence. *The Future of Environmental Criticism: Environmental Crisis and Literary Imagination.* Blackwell, 2005.

Bulhof, Francis. "Agendath Again." *JJQ* 7.4 (Summer 1970): 326–332.

Burchardt, Jeremy. *Paradise Lost: Rural Idyll and Social Change since 1800.* London: I.B. Taurus, 2002.

Byrnes, Robert. "Agendath Netaim Discovered: Why Bloom Isn't a Zionist." *JJQ* 29.4 (Summer 1992): 833–838.

Callanan, Frank. "James Joyce and *The United Irishman*, Paris 1902-3." *Dublin James Joyce Journal* 3 (2010): 51–103.

Campbell, John Francis. Friendly Societies in Ireland 1850-1960: with particular reference to Ancient Order of Hibernians and the Irish

National Foresters. Unpublished M. Litt. Thesis., Trinity College Dublin, October 1998.

Caraher, Brian G. "A 'ruin of all space, shattered glass and toppling masonry': Joyce's orientalism in the context of 11 September 2001 and 1922." *Textual Practice* 18(4), 2004. 497–520.

Caraher, Brian G . "Cultural Politics and the Reading of 'Joyce': Cultural Semiotics, Socialism, Irish Autonomy, and 'Scritti Italiani.'" *JJQ* 36.2 (1999): 171–214.

Caraher, Brian G . "A Question of Genre: Generic Experimentation, Self-composition, and the Problem of Egoism in *Ulysses*." *ELH* 54.1 (Spring 1987): 183–214.

Cheng, Vincent J. *Joyce, Race, and Empire*. Cambridge: Cambridge University Press, 1995.

Clark, Timothy. *The Cambridge Introduction to Literature and the Environment*. Cambridge: Cambridge University Press, 2011.

Cleary, Joe. "Introduction: Ireland and Modernity." *The Cambridge Companion to Modern Irish Culture*, ed. Joe Cleary and Claire Connolly. Cambridge: Cambridge University Press, 2005. 1–21.

Conley, Tim. "Eco-Joyce: The Environmental Imagination of James Joyce." Book Review. *Studies Irlandese* (2015) 10: 135–170. <http://www.estudiosirlandeses.org/reviews/eco-joyce-the-environmental-imagination-of-james-joyce/> Accessed 15 March 2016.

Connolly, James. *James Connolly: Selected Writings*. Ed. Peter Berresford Ellis. London: Pluto Press, 1997.

Creese, Walter L. *The Search for Environment: the Garden City Before and After*. New Haven: Yale University Press, 1966.

Curran, C. P. *Under the Receding Wave*. London: Gill and Macmillan, 1970.

Cusick, Christine, ed. *Out of the Earth: Ecocritical Readings of Irish Texts*. Cork: Cork University Press, 2010.

Davis, Cynthia. "The Garden and Resistance in Diasporic Literature: an Ecocritical Approach." *Landscape and Empire: 1770—2000*. Ed. Glenn Hooper. Vermont: Ashgate, 2004. 195–206.

Davison, Neil R. "'Still an Idea Behind It': Trieste, Jewishness, and Zionism in '*Ulysses*'" *JJQ* 38.3/4 (Spring-Summer 2001): 373–394.

Davison, Neil R . "Why Bloom is Not 'Frum', or Jewishness and Postcolonialism in '*Ulysses*.'" *JJQ* 39.4 (Summer 2002): 679–716.

DeLoughrey, Elizabeth, and George B. Handley, eds. *Postcolonial Ecologies: Literatures of the Environment.* Oxford: Oxford University Press, 2011.

Donaghy, Henry J. *James Clarence Mangan.* New York: Twayne Publishers, 1974.

Duffy, Enda. "Disappearing Dublin: *Ulysses*, Postcoloniality, and the Politics of Space." *Semicolonial Joyce.* Ed. Derek Attridge and Marjorie Howes. Cambridge: Cambridge University Press, 2000. 37–57.

Eagleton, Terry. *Heathcliff and the Great Hunger: Studies in Irish Culture.* Verso, 1995.

Ellmann, Richard. *James Joyce.* New and revised edition. Oxford: Oxford University Press, 1982.

Ellmann, Richard . *The Consciousness of Joyce.* London: Faber and Faber, 1977.

Engels, Friedrich. *The Housing Question.* Mascow: Progressive Publishers, 1970.

Fairhall, James. "Northsiders." *Joyce: Feminism/Post/Colonialism.* Ed. by Ellen Carol Jones. Amsterdam: Rodopi, 1998. pp. 43–80.

Fairhall, James . *James Joyce and the Question of History.* Cambridge: Cambridge University Press, 1993.

Fisch, Max Harold, and Thomas Goddard Bergin. "Introduction." *The Autobiography of Giambattista Vico.* Trans. by Max Harold Fisch and Thomas Goddard Bergin. Ithaca: Cornell University Press, 1944. 1–108.

Fogarty, Anne. "Foreword." *Eco-Joyce: the Environmental Imagination of James Joyce.* Eds. by Robert Brazeau and Derek Gladwin. Cork: Cork University Press, 2014. pp. xv–xviii.

Forbes, A. C. "Tree Planting in Ireland during Four Centuries." *Proceedings of the Royal Irish Academy. Section C: Archaeology, Celtic Studies, History, Linguistics, Literature* 41 (1932–1934): 168–199.

Foster, John Wilson, ed. *Nature in Ireland: A Scientific and Cultural History.* Dublin: Lilliput Press, 1997.

Foucault, Michel, and Jay Miskowiec. "Of Other Spaces." *Diacritics* 16:1 (Spring 1986): 22–27.

Fraser, Murray. *John Bull's Other Homes: State Housing and British Policy in Ireland, 1883-1922.* Liverpool: Liverpool University Press, 1996.

Garrard, Greg. *Ecocriticism.* New York: Routledge, 2004.

Garrard, Greg , ed. *The Oxford Handbook of Ecocriticism.* Oxford: Oxford University Press, 2014.

Geddes, Patrick, and F. C. Mears. *City and Town Planning Exhibition: Guide-Book and Outline Catalogue.* Dublin: Browne and Nolan, 1911.

Gibson, Andrew. *Joyce's Revenge: History, Politics, and Aesthetics in* Ulysses. Oxford: Oxford University Press, 2002.

Gibson, Andrew . "Introduction." *Joyce's Ithaca*. Amsterdam: Rodopi, 1996. 3–27.

Gifford, Don, and Robert J. Seidman. *Ulysses Annotated: Notes for James Joyce's Ulysses*. Berkeley: University of California Press, 1974.

Gilbert, Stuart. *James Joyce's 'Ulysses': A study*. London: Penguin, 1963.

Glotfelty, Cheryll, and Harold Fromm, eds. *The Ecocriticism Reader: Landmarks in Literary Ecology*. University of Georgia Press, 1996.

Graham, P. Anderson. *Reclaiming the Waste: Britain's Most Urgent Problem*. London: Country Life Library, 1916.

Griffith, Arthur. *The Resurrection of Hungary: A Parallel for Ireland*. (1904). With introduction by Patrick Murray. Dublin: University College Dublin Press, 2003.

Groden, Michael. "Joyce at Work on 'Cyclops': Toward a Biography of *Ulysses*." *James Joyce Quarterly* 44.2 (2007): 217–45.

Groden, Michael . "The National Library of Ireland's New Joyce Manuscripts." *Joyce in Trieste: an Album of Risky Readings*. Eds. by Sebastian D. G. Knowles, Geert Lernout, and John McCourt. University Press of Florida, 2007.

Groden, Michael . *Ulysses in Progress*. Princeton: Princeton University Press, 1977.

Groden, Michael . "'Cyclops' in Progress, 1919." *JJQ*, Vol. 12, No. 1/2, Textual Studies Issue (Fall 1974 – Winter 1975): 123–68.

Hadfield, Andrew, and Willy Maley. Introduction. *A View of the State of Ireland: from the first printed edition (1633)*. By Edmund Spenser. Eds. by Andrew Hadfielf and Willy Maley. Oxford: Blackwell Publishers, 1997.

Hall, Pater, Dennis Hardy and Colin Ward. "Commentators' Introduction." *To-morrow: a Peaceful Path to Real Reform*. By Ebenezer Howard. With commentary by Pater Hall, Dennis Hardy and Colin Ward. London: Routledge, 2003.

Harrison, Robert Pogue. *Gardens: an Essay on the Human Condition*. Chicago: University of Chicago Press, 2008.

Hart, Clive. "Wandering Rocks." *James Joyce's "Ulysses": Critical Essays*. Ed. by Clive Hart and David Hayman. Berkeley: University of California Press, 1974. 181–216.

Hawkins, Gay. *The Ethics of Waste: How We Relate to Rubbish*. Rowman & Littlefield, 2006.

Herr, Cheryl Temple. "Joyce and the Everynight." *Eco-Joyce: the Environmental Imagination of James Joyce*. Cork: Cork University Press, 2014. 38–58.

Hiltner, Ken, ed. *Ecocriticism: The Essential Reader*. London: Rutledge, 2015.

Hofheinz, Thomas C. *Joyce and the Invention of Irish History: Finnegans Wake in context*. Cambridge University Press, 1995.

Hone, Joseph. *W. B. Yeats, 1865-1939*. London: Mac millan & Co, 1943.

Howard, Ebenezer. *To-morrow: a Peaceful Path to Real Reform*. With commentary by Pater Hall, Dennis Hardy and Colin Ward. London: Routledge, 2003.

Howes, Marjorie. "'Goodbye Ireland I'm going to Gort': Geography, Scale, and Narrating the Nation." *Semicolonial Joyce*. Ed. Derek Attridge and Marjorie Howes. Cambridge: Cambridge University Press, 2000. 58–77.

Hoyle, Martin. *The Story of Gardening*. London: Journeyman Press, 1991.

Huggan, Graham, and Helen Tiffin. *Postcolonial Ecocriticism: Literature, Animals, Environment*. London, New York: Rutledge, 2010.

Hutton, Clare. "Joyce, the Library Episode, and the Institutions of Revivalism." *Joyce, Ireland, Britain*. Ed. by Andrew Gibson and Len Platt. Gainesville: University Press of Florida, 2006. 122–139.

Hutton, Clare . "Joyce and the Institutions of Revivalism." *Irish University Review* 33.1 (Spring/Summer 2003): 117–32.

Hyamson, Albert M. "Anti-Semitism." *Zionism and the Jewish Future*. Ed. by H. Sacher. London: John Murray, 1917. 59–86.

Hyde, Douglas. *Beside the Fire: a Collection of Irish Gaelic Folk Stories*. 1910.

Jarman, Neil. *Material Conflicts: Parades and Visual Displays in Northern Ireland*. Oxford: Berge, 1997.

Jones, Ellen Carol. "Commodious Recirculation: Commodity and Dream in Joyce's *Ulysses*," *Joyce and Advertising*, eds. Garry Leonard and Jennifer Wicke, Special Issue, *James Joyce Quarterly* 30.4–31.1 (Summer/Fall 1993). 739–56.

Kelly, Fergus. "The Old Irish Tree List." *Celtica* 11 (1976): 107–24.

Kelly, Joseph. *Our Joyce: from Outcast to Icon*. Austin: University of Texas Press, 1998.

Kershner, R.B. Jr. "The Reverend J.L. Porter: Bloom's Guide to the East." *JJQ* 24.3 (Spring 1987): 365–67.

Kiberd, Declan. Ulysses *and Us: the Art of Everyday Living*. Faber and Faber, 2009.

Kilroy, James. *James Clarence Mangan*. Lewisburg: Bucknell University Press, 1970.

Kropotkin, Peter. *The Conquest of Bread*. London: Chapman & Hall, 1913.

Kropotkin, Peter . *Fields, Factories, and Workshops; or, Industry combined with agriculture and brain work with manual work*. London: Swan Sonnenschein & Co., 1907.

Kupinse, William. "Scrupulous Greenness: Eco-Joyce: The Environmental Imagination of James Joyce" Book Review. *BREAC: A Digital Journal of Irish Studies*. University of Notre Dame. April 2015. <http://breac.nd.edu/articles/56844-scrupulous-greenness/> Accessed 16 March 2016.

Lacivita, Alison. *"Eco-Joyce: The Environmental Imagination of James Joyce."* *Green Letters* 19.2 (2015): 210–212.

Lane, Leeann. "'It is In the Cottages and Farmers' Houses that the Nation is Born': AE's *Irish Homestead* and the Cultural Revival." *Irish University Review* 33.1 (Spring/Summer 2003): 165–81.

Laporte, Dominique. *History of Shit*. Trans. by Nadia Benabid and Rodolphe el-Khoury. Cambridge: MIT Press, 2000.

Latham, Sean. "Twenty-first-century critical contexts." *James Joyce in Context*. Ed. John McCourt. Cambridge: Cambridge University Press, 2009. pp. 148–59.

Lee, Joseph. *Ireland: 1912-1985*. Cambridge: Cambridge University Press, 1989.

Leonard, Garry. "Joyce and Advertising: Advertising and Commodity Culture in Joyce's Fiction." *James Joyce Quarterly*, Vol. 30/31, 30:4 – 31:1, Joyce and Advertising (Summer – Fall, 1993): 573–92.

Litz, A. Walton. "Vico and Joyce." *Giambattista Vico: an International Symposium*. Ed. by Giorgio Tagliacozzo and Hayden V. White. Baltimore: the Johns Hopkins Press, 1969. 245–55.

Lloyd, David. *Irish Times: Temporalities of Modernity*. Dublin: Field Day, 2008.

Long, Gerald. "William Rooney's 'An Irish Rural Library.'" *Dublin James Joyce Journal* 3 (2010): 104–13.

Lousley, Cheryl. "Ecocriticism and the Politics of Representation." *The Oxford Handbook of Ecocriticism*. Ed. by Greg Garrard. Oxford: Oxford University Press, 2014. pp. 155–171.

Lyons, F.S.L. *Ireland Since the Famine*. Fontana Press, 1985.

Lysaght, Meadhbh. "Trinity College Schools' Competition Junior Gold Medal Winner: Dublin's Wholesale Fruit & Vegetable Market." *History Ireland* 4:3 (Autumn 1993): 42–45.

MacCabe, Colin. *James Joyce and the Revolution of the Word*. 2nd ed. Palgrave, 2002.

MacCannell, Dean. "Landscaping the Unconscious." *The Meaning of Gardens: Idea, Place, and Action*. Ed. by Mark Francis and Randolph T. Hester, Jr. Cambridge: MIT Press, 1990. 94–101.

MacDonagh, Oliver. *States of Mind: a Study of Anglo-Irish Conflict, 1780-1980*. London: Allen & Unwin, 1983.

Mali, Joseph. "Mythology and Counter-history: the New Critical Art of Vico and Joyce." *Vico and Joyce*, Donald Phillip Verene, ed. Buffalo: SUNY Press, 1987. 32–47.

Mangan, James Clarence. *Selected Writings*. Ed. and intro. by Sean Ryder. Dublin: University College Dublin Press, 2004.

Manganiello, Dominic. "Vico's Ideal History and Joyce's Language." *Vico and Joyce*, Donald Phillip Verene, ed. Buffalo: SUNY Press, 1987. 196–206.

Manganiello, Dominic. *Joyce's Politics*. London: Routledge & Kegan, 1980.

Marx, Karl. *Capital: A Critique of Political Economy. Vol. II: The Process of Circulation of Capital*. Ed. by Frederick Engels. London: Lawrence and Wishart, 1974.

Marx, Leo. *The Machine in the Garden: Technology and the Pastoral Ideal in America*. London: Oxford University Press, 1964.

Mathews, P. J. "'A.E.I.O.U.': Joyce and the *Irish Homestead*." *Joyce on the Threshold*. Ed. Anne Fogarty and Timothy Martin. Gainesville: University Press of Florida, 2005. 151–68.

Mathews, P. J. *Revival: the Abbey Theatre, Sinn Féin, the Gaelic League and the Co-operative Movement*. Cork: Cork University Press, 2003.

McAteer, Michael. *Standish O'Grady, AE and Yeats: History, Politics, Culture*. Dublin: Irish Academic Press, 2002.

McCracken, Eileen. *The Irish Woods Since Tudor Times: Distribution and Exploitation*. Newton; Abbott: David & Charles, 1971.

McDowell, Michael J. "The Bakhtinian Road to Ecological Insight." *The Ecocriticism Reader: Landmarks in Literary Ecology*. Ed. Cheryll Glotfelty and Harold Fromm. Athens: University of Georgia Press, 1996. 371–91.

McGee, Owen. "Power, John Wyse." *Dictionary of Irish Biography*. (Eds.)James Mcguire, James Quinn. Cambridge, United Kingdom: Cambridge University Press, 2009. <http://dib.cambridge.org/viewReadPage.do?articleId=a7469> Accessed 5 March 2010.

McKay, George. *Radical Gardening: Politics, Idealism & Rebellion in the Garden*. London: Frances Lincoln, 2011.

Meacham, Standish. *Regaining Paradise: Englishness and the Early Garden City Movement*. New Haven: Yale University Press, 1998.

Morris, Ewan. *Our Own Devices: National Symbols and Political Conflict in Twentieth-Century Ireland*. Dublin: Irish Academic Press, 2005.

Morton, Timothy. *The Ecological Thought*. Cambridge: Harvard University Press, 2010.

Morton, Timothy . *Ecology without Nature: Rethinking Environmental Aesthetics*. Cambridge: Harvard University Press, 2007.

Mumford, Lewis. "The Garden City Idea and Modern Planning." *Garden Cities of To-morrow*. By Ebenezer Howard. Ed. by F. J. Osborn. London: Faber and Faber, 1945.

Murphy, William, and Lesa Ní Mhunghaile. "Power, Jennie Wyse." *Dictionary of Irish Biography*. (Eds.)James Mcguire, James Quinn. Cambridge, United Kingdom: Cambridge University Press, 2009. <http://dib.cambridge.org/viewReadPage.do?articleId=a7454> Accessed 01 March 2010.

Nash, John. *James Joyce and the Act of Reception: Reading, Ireland, Modernism*. Cambridge: Cambridge University Press, 2006.

Neeson, Eoin. "Woodland in History and Culture." *Nature in Ireland: A Scientific and Cultural History*. Ed. by John Wilson Foster. Dublin: Lilliput Press, 1997. 133–56.

Neeson, Eoin. *A History of Irish Forestry*. Dublin: Lilliput Press, 1991.

Nolan, Emer. *James Joyce and Nationalism*. London: Routledge, 1995.

O'Brien, Joseph V. *Dear, Dirty Dublin: a City in Distress, 1899-1916*. University of California Press, 1992.

O'Callaghan, Katherine. "Eco-Joyce: the Environmental Imagination of James Joyce." *Irish Studies Review* 23.4 (2015): 508–510.

Ó Gráda, Cormac. *Jewish Ireland in the Age of Joyce: a Socioeconomic History*. Princeton: Princeton University Press, 2006.

Osteen, Mark. "Seeking Renewal: Bloom, Advertising, and the Domestic Economy." *Joyce and Advertising*, ed. Garry Leonard and Jennifer Wicke, Special Issue, *James Joyce Quarterly* 30.4–31.1 (Summer/Fall 1993). 717–38.

Parish, Charles. "Agenbite of Agendath Netaim." *JJQ* 6.3 (Spring 1969): 237–241.

Pearson, Peter. *The Heart of Dublin: Resurgence of an Historic City*. Dublin: The O'Brien Press, 2000.

Platt, Len. *Joyce and the Anglo-Irish: a Study of Joyce and the Literary Revival*. Amsterdam: Rodopi, 1998.

Platt, Len. "'If Brian Boru Could But Come Back And See Old Dublin Now': Materialism, the National Culture And *Ulysses* 17." *Joyce's Ithaca*. Ed. by Andrew Gibson. Amsterdam: Rodopi, 1996. 105-32.

Plock, Vike Martina. "A Feat of Strength in 'Ithaca': Eugen Sandow and Physical Culture in Joyce's *Ulysses*." *Journal of Modern Literature* 30.1 (2006). 129-39.

Pollard, Captain H. B. C. *The Secret Societies of Ireland: Their Rise and Progress*. London: Philip Allan & Co., 1922.

Pye, Gillian. "Introduction: Trash as Cultural Category." *Trash Culture: Objects and Obsolence in Cultural Perspective*. Ed. by Gillian Pye. Vol. 11 of *Cultural Interactions: Studies in the Relationship between the Arts*. Bern: Peter Lang, 2010.

Reizbaum, Marilyn. "Swiss Customs: Zurich's Sources for Joyce's Judaica." *JJQ* 27.2 (Winter 1990): 203-218.

Reynolds, Mary T. "The City in Vico, Dante, and Joyce." *Vico and Joyce*. Donald Phillip Verene, ed. Buffalo: SUNY Press, 1987. 110-22.

Riley, Robert B. "Flower, Power, and Sex." *The Meaning of Gardens: Idea, Place, and Action*. Ed. by Mark Francis and Randolph T. Hester, Jr. Cambridge: MIT Press, 1990. 60-75.

Robichaud, Paul. "Joyce, Vico, and National Narrative." *JJQ* 41.1/2, Post-Industrial Joyce (Fall 2003-Winter 2004): 185-96.

Roche, Desmond. *Local Government in Ireland*. Dublin: Institute of Public Administration, 1982.

Rose, Danis. "The Source of Mr. Bloom's Wealth." *James Joyce Quarterly* 25.1 (Fall 1987): 128-32.

Rubenstein, Michael. *Public Works: Infrastructure, Irish Modernism, and the Postcolonial*. Notre Dame: University of Notre Dame Press, 2010.

Rueckert, William. "Literature and Ecology." *The Ecocriticism Reader: Landmarks in Literary Ecology*. Ed. by Cheryl Glotfelty and Harold Fromm. Athens: University of Georgia Press, 1996. pp. 105-123.

Russell, George W. (A.E.). *Selections from the Contributions to the Irish Homestead*. Vol. 1-2. Ed. By Henry Summerfield. London: Colin Smythe, 1978.

Russell, George W. *Co-operation and Nationality: A Guide for Rural Reformers from This to the Next Generation*. Dublin: Maunsel & Company, 1912.

Ryder, Sean. Ed. "Introduction." *James Clarence Mangan: Selected Writings*. Dublin: University College Dublin Press, 2004.

Ryle, Martin. "Eco-Joyce: The Environmental Imagination of James Joyce." Book Review. *Textual Practice* (2014) 28.6: 1153-57.

Sacher, H. "A Century of Jewish History." *Zionism and the Jewish Future*. Ed. by H. Sacher. London: John Murray, 1917. 12–58.

Sales, Roger. *English Literature in History 1780-1830: Pastoral and Politics*. London: Hutchinson, 1983.

Sandquist, Brigitte L. "The Tree Wedding in 'Cyclops' and the Ramifications of Cata-logic." *James Joyce Quarterly* 33.2 (Winter 1996): 195–209.

Scanlan, John. *On Garbage*. Bath: Reaktion Books, 2005.

Schiller, Friedrich. "On Naive and Sentimental Poetry", trans. Julius A. Elias, in *German Aesthetic and Literary Criticism*. Urbana and Chicago: University of Illinois Press, 1985.

Schutte, William M. "An Index of Recurrent Elements in 'Ulysses': 'Cyclops.'" *JJQ*, Vol. 16, No. 1/2, Structuralist/Reader Response Issue (Fall, 1978 – Winter, 1979): 161–80.

Schwartzman, Myron. "The V.A.8 Copybook: An Early Draft of the "Cyclops" Chapter of "Ulysses" with Notes on Its Development." *JJQ*, Vol. 12, No. 1/2, Textual Studies Issue (Fall 1974 – Winter 1975): 64–122.

Seidel, Michael. *Epic Geography: James Joyce's Ulysses*. New Jersey: Princeton University Press, 1976.

Senn, Fritz. "Ovidian Roots of Gigantism in Joyce's 'Ulysses.'" *Journal of Modern Literature* 14.4 (Spring 1989): 561–77.

Senn, Fritz . "'Trivia Ulysseana' IV." *James Joyce Quarterly* 19.2 (Winter 1982): 151–78.

Sheehy, Jeanne. *The Rediscovery of Ireland's Past: the Celtic Revival 1830-1930*. London: Thames and Hudson, 1980.

Smyth, Gerry. "'Shite and Sheep': An Ecocritical Perspective on Two Recent Irish Novels" in special issue of *Contemporary Irish Fiction, Irish Universeity Review* 30:1 (2000), pp. 163–78.

Smyth, Gerry . *Space and Irish Cultural Imagination*. New York: Palgrave, 2001.

Soper, Kate. *What is Nature?: Culture, Politics and the Non-Human*. Blackwell, 1995.

Spenser, Edmund. *A View of the State of Ireland: from the first printed edition* (1633). Eds. by Andrew Hadfielf and Willy Maley. Oxford: Blackwell Publishers, 1997.

Spoo, Robert. *James Joyce and the Language of History: Dedalus's Nightmare*. Oxford: Oxford University Press, 1994.

Spurr, David. "Colonial Spaces in Joyce's Dublin." *James Joyce Quarterly* 37:1/2, Dublin and Dubliners (Fall 1999 – Winter 2000): 23–42.

Swenarton, Mark. *Building the New Jerusalem: Architecture, Housing and Politics 1900-1930*. Bracknell: IHS BRE Press, 2008.

Symons, Arthur. *The Symbolist Movement in Literature*. Ed. Richard Ellmann. Kessinger Publishing, 2004.

Tolkowsky, S. "The Jews and the Economic Development of Palestine. (With Map.)" *Zionism and the Jewish Future*. Ed. by H. Sacher. London: John Murray, 1917. 138–170.

Turner, Michael. *After the Famine: Irish agriculture, 1850-1914*. Cambridge: Cambridge University Press, 1996.

Vico, Giambattista. *The New Science of Giambattista Vico: unabridged translation of the third edition (1744) with the addition of "Practic of the New Science."* Trans. by Thomas Goddard Bergin and Max Harold Fisch. Ithaca: Cornell University Press, 1984.

Ware, Rob. "*Eco-Joyce*: The Environmental Imagination of James Joyce." Book Review. *Interdisciplinary Study of Literature and Environment* (2014) 21.4: 930–32.

Weizmann, Dr. Chaim. "Introduction." *Zionism and the Jewish Future*. Ed. by H. Sacher. London: John Murray, 1917. 1–11.

Wells, H. G. "James Joyce." *New Republic* 10 (10 March 1917). 159.

Wenzell, Tim. *Emerald Green: An Ecocritical Study of Irish Literature*. Newcastle: Cambridge Scholars Publishing, 2009.

Whiteley, Gillian. *Junk: Art and Politics of Trash*. London: I.B. Tauris & Co., 2011.

Wicht, Wolfgang. "'Bleibtreustrasse 34, Berlin, W. 15.' (U 4.199), Once Again." *JJQ* 40.4 (Summer 2003): 797–810.

Wicke, Jennifer. "Joyce and Consumer Culture." *The Cambridge Companion to James Joyce*. Ed. Derek Attridge. 2nd ed. Cambridge University Press, 2004. 234–53.

Wicke, Jennifer. "Modernity Must Advertise: Aura, Desire, and Decolonization in Joyce." *James Joyce Quarterly* 30:4 - 31:1, no. 1, Joyce and Advertising (Summer - Fall 1993): 593–613.

Wilding, Paul. "The Housing and Town Planning Act 1919 -- a Study in the Making of Social Policy." *Journal of Social Policy* 2.4 (1973): 317–34.

Williams, Edwin W. "Agendath Netaim: Promised Land or Waste Land." *MFS Modern Fiction Studies* 32.2 (Summer 1986): 228–235.

Williams, Raymond. *The Country and the City*. New York: Oxford University Press, 1973.

Index

A
"Agendath Netaim" 23, 70, 74, 79
Abbey Theatre 37
Abercrombie, Patrick 49, 55, 67
Agricultural and Technical
  Instruction, Dept of
  (DATI) 104–105, 137
agriculture 35, 63, 43–44, 49, 61, 70,
  101, 104, 137, 109, 143
Albion, Robert Greenhaugh 103
Ancient Order of Foresters
  (AOF) 112, 138
Anglo-Irish Ascendency *see*
  Ascendency 38, 40
Anglo-Irish Revival 64, 149, 163
Arnold, Matthew 83
Arts and Crafts
  Movement 47, 65, 59
Ascendency 39–40, 44

B
Bakhtin, M. M. 141, 169, 161, 163
Balfour Declaration 77–79
Barnacle, Nora 46, 129
Beckett, Samuel 124
Bloom, Leopold 21–22, 62, 31,
  33–34, 36–37, 39–48, 50–57,
  59–61, 69–74, 96, 77–80, 82–83,
  86, 88, 90–94, 112, 138, 117, 121,
  125, 129, 133, 150, 154, 167, 161
– "New Bloomusalem" 34,
  50–51, 61, 77
– as a gardener in "Ithaca" 60
bog 108, 131–132
bogland *see* bog 101, 136, 108
Bookchin, Murray 59
Bosinelli, Rosa Maria 125, 136, 140
Brehon Law 23, 104, 131–132, 134

British imperialism 89
Budgen, Frank 153
Buell, Lawrence 18

C
Campbell, John F. 138, 113, 139, 117
capitalism 33, 41, 44, 51, 58, 61, 84,
  127, 163
Caraher, Brian 9, 32, 62, 77–78
Celtic Revival 41, 65, 47, 128, 150
Cheng, Vincent 31, 63, 150, 166
civilisation 22, 70, 72, 83, 85, 87,
  127, 136, 142
Cleary, Joe 84, 89
colonialism 25, 72, 90, 98,
  134–135, 146
– in Ireland 16–18, 23–24, 33,
  35–36, 64, 39–40, 42, 46, 65, 55,
  57, 73–74, 95, 97, 84, 136, 101,
  103–105, 137, 107–110, 113,
  133–134, 143–145, 158, 171
Connolly, James 133–134
consumption 20–22, 24, 26, 34,
  71–73, 80–81, 85, 89–94, 108, 151,
  156, 159, 171–172
– commodity 45, 57, 59, 80, 90,
  156–159
– commodity culture 21, 33, 59, 159
– in "Ithaca" 22, 32–33, 35,
  41, 43, 55
Co-operative movement 22, 33, 46
Creese, Walter 58, 61, 68
cultural nationalism 22, 46, 75, 87
Cumann na mBan 119, 121–122,
  128, 131, 133
Curran, C. P. 117
Cusack, Michael 121, 150
Cusick, Christine 19

D
Davison, Neil R. 96, 75, 77
Davitt, Michael 36, 53
Dedalus, Stephen 21, 37, 80, 83, 85–86, 88, 93, 150
Douglas, Mary 71
Dubliners (Joyce) 37, 88, 98, 160
Duffy, Enda 159, 168

E
Eagleton, Terry 16, 143, 164, 144
Easter Rising 55, 114, 128
Ecocriticisim Reader, The (Glotfelty & Fromm) 17
Ecocriticism 17–20, 25, 163, 171
– and Irish studies 20
– *Ecocriticism Reader, The* 17
– Joyce and 19
Eco-Joyce: *The Environmental Imagination of James Joyce* (Brazeau & Gladwin) 9, 95
Eglinton, John 32
Ellmann, Richard 7, 13, 96, 137, 109
Epicurus 60
Evening Telegraph 121, 151

F
Fairhall, James 31, 33, 37, 137
famine 26, 16, 35, 63, 84–85, 143, 164
– The Great Famine 16, 84–85
Ferguson, Samuel 41, 147
feudalism 122, 126–127, 131, 134
Finnegans Wake (Joyce) 13, 19–20, 69, 86, 123–126, 134–135
Fogarty, Anne 9, 25, 62
Forbes, A. C. 101, 106, 137
forest 23, 136, 103–104, 106, 112, 128, 134–135, 150
– afforestation 23–24, 74, 95, 97, 101, 104–105, 107–111, 122, 126–127, 131, 135, 171
– deforestation 24, 104–105
– reafforestation 69, 73, 78, 105
– Vico and 131, 135
Foster, John Wilson 26, 17
Foucault, Michel 159, 168, 160
Freeman's Journal, The 83, 107, 113, 115, 121, 153–154, 167
friendly society 113, 133

G
Gaelic Athlete Association (GAA) 121
Gaelic League 33, 37, 63, 41, 107, 113, 121
garbage *see* waste 20, 71–72, 79, 82, 88–89, 91–92, 94, 171
garden 21–22, 44, 46–52, 54, 57–61, 72, 92–93, 171–172
– "Bloom of Flowerville" 21, 33, 39–40, 43, 51, 61
– gardening 25, 62, 41, 44, 58–60, 68, 61, 72–73, 98, 92
Garden City Movement 21, 48, 50, 58, 61–62
Garrard, Greg 18
Geddes, Patrick 49, 55, 66, 57
Gibson, Andrew 62, 38, 64, 39, 41, 87, 97, 137, 109, 138, 111–112, 140, 147, 165, 148–150, 152, 159, 168
Gifford, Don 35, 63, 95, 74, 80, 96, 85, 97, 86, 98, 137, 103, 164–166, 151, 168
Gilbert, Stuart 13, 123–124
Glotfelty, Cheryll 17
Graham, P. Anderson 69, 73–74, 90
Gregory, Lady Augusta 38, 63, 147
Griffith, Arthur 42, 107–109, 137, 110–112, 116–117, 121, 127
– *Resurrection of Hungary* 109
– *Sinn Féin Policy* 42, 109–110

H

Harrison, Robert Pogue 23, 57, 68, 62, 172
Hart, Clive 88, 98
Hawkins, Gay 71-72, 95, 81, 96, 88, 97, 89, 98, 91-92
Herr, Cheryl 19, 70, 95, 88
Hiltner, Ken 25, 27
history 16, 26, 17, 19-23, 25, 31, 63, 36, 40, 46, 65, 48, 54-55, 67, 57, 62, 70-72, 80-82, 97, 83-84, 89, 98, 101-102, 107, 109, 124-126, 131-132, 135, 144, 147-149, 152, 160, 171-172
- and the environment 9, 17
- of Irish forestry 101, 104-105, 132
Hofheinz, Thomas C. 125, 133, 140
Home Rule 16, 22, 34, 36, 38, 48, 54, 57, 108, 161
Homer 123-124, 126-127, 164, 143
Housing and Town Planning Act 54
Housing and Town Planning Association of Ireland (HTPAI) 54
Howard, Ebenezer 21, 48-50, 66, 51-55, 58-59, 62
- *Garden Cities of To-morrow* 21, 66, 50-53, 62
Howes, Marjories 146
Hoyles, Martin 61
Hutchinson, John 39, 64, 66, 112, 138
Hutton, Clare 32, 62, 38, 64
Hyde, Douglas 38, 121, 147

I

Igoe, Vivien 46
imperialism *see* colonialism, British imperialism 72, 81, 83, 85, 107, 149
Irish Agricultural Organisation Society (IAOS) 33, 37, 104

Irish Homestead, The 21-22, 32-33, 37, 39-43, 45, 48, 66, 49, 51, 62, 172
Irish National Foresters (INF) 106, 112, 138, 113-114, 117, 119-121, 128, 133, 135
Irish Republican Brotherhood (IRB) 121, 128

J

Jarman, Neil 114, 138-139, 120
Jefferson, Thomas 61
Jones, Ellen Carol 34, 45, 65
Joyce, James 7, 9, 13, 19, 21, 62, 46, 69, 87, 136, 101, 137, 110, 122, 134, 144, 172
Joyce, Stanislaus 110

K

Kiberd, Declan 32, 93, 98
Kropotkin, Peter 42-44, 65, 45, 49-50

L

land reform 16, 26, 34-36, 53
landscape 15-16, 18, 24, 40, 44, 48, 58-59, 101, 104, 127, 130, 150, 163, 171-172
- idyllic 24, 41, 48, 58, 61-62, 142-143, 145-146, 151, 163, 171
- Irish landscape 40, 101, 104
Lane, Leeanne 39, 64, 152, 155
Laporte, Dominique 70, 80-81, 96, 82, 85, 97, 86-87, 98
Latham, Sean 19
Lee, Joseph 35
Leonard, Garry 19, 156
Literary Revival, The, see *Celtic Revival* 32-33, 64
Little Review, The 24, 87, 132, 149
Lloyd, David 84, 97
Lyons, F.S.L. 26, 63, 37

M

MacCannell, Dean 60
MacDonagh, Oliver 15
Mangan, James Clarence 24, 141, 143, 164, 144–145, 147–149, 157–159
Manganiello, Dominic 65, 43, 137, 109, 138
market *see* marketplace 22, 24, 44, 48, 84, 106–107, 148, 153–155, 167–168, 156–161
marketplace 23–24, 59, 159, 163, 171
– as a colonial space 159
Marx, Leo 59
Mathews, P. J. 37, 63, 38, 64, 39
McCracken, Eileen 101–102, 136
McKay, George 68, 60
modernism 20, 65, 94, 159, 171
– and advertisement 37
– and consumption 21–22, 24, 26, 73, 81, 89, 172
– and the environment 9, 17
– colonial modernism 94
modernity 26, 19–20, 22, 41, 80, 88, 90, 94, 143, 153, 157, 159–160
– colonial modernity 26, 159
– cultural modernity 153, 157
Moore, George 37
Moore, Thomas 165
Moran, D. P. 41, 116
Morris, William 49, 58–59, 164
Morton, Timothy 19

N

Nash, John 32, 62
nation 23, 25–26, 37, 54, 58, 61, 95, 109–110, 123, 141, 143–144, 146, 149, 172
– and nature 18, 60, 135, 172
nationalism 16, 22, 36–39, 48, 55, 57, 62, 75, 77, 80, 85, 87, 107, 109, 114, 122, 127, 133, 164, 141, 144–145, 149, 154, 171
– constitutional nationalism 22, 37–39, 48
– cultural nationalism 22, 46, 75, 87
– Irish nationalism 16, 36–38, 57, 77, 107, 122, 127, 133, 145, 149, 154
Neeson, Eoin 136, 101, 137, 104–105, 132, 140
Nolan, Emer 116, 135, 141

O

O'Donnell of the Glens *see* O'Donnell, Hugh Roe 104
O'Donnell, Hugh Roe 104, 133
O'Grady, Standish 41, 112, 147
Occasional, Critical, and Political Writings (Joyce) 13, 136
– "Home Rule Comes of Age" 161
Ovid 122, 130, 151

P

Palestine 69–70, 74–79
parade 94, 114–117
Parnell, Charles Stewart 16, 36, 38, 63, 50, 113, 115
pastoral 15, 24–25, 141–146, 151, 157, 161, 163, 172
Pearson, Peter 152, 166, 153
Petrie, George 40, 64, 41
picturesque, the 24, 115–117, 142–143
Plantation, of Ulster 46, 104, 132
Plato 60
Platt, Len 62, 38, 64, 39
Plunkett, Horace 37, 40, 49, 57, 104, 107
pollution 81
Portrait of the Artist as a Young Man, A (Joyce) 87–88
public works 95, 82

R

River Liffey 73, 91
Rubenstein, Michael 95, 97, 82, 84, 98, 89, 94
Rueckert, William 17
rurality 15–16, 37, 41–42
Russell, George (AE) 32, 37–38, 42, 66, 113
Ryder, Sean 144, 164, 145

S

Sales, Roger 142
Sandquist, Brigitte L. 122, 125, 140, 127, 130–131, 134, 164
Scanlan, John 82
Senn, Fritz 114, 122, 130, 140, 155
sewage 22, 25, 69–70, 72, 95, 80–81, 83, 85–94, 171–172
– and power 50, 70, 86–87, 93
– sewer of consciousness 23, 70, 89, 91, 94
Sheehy, Jeanne 65–66, 48
Sinn Féin 42, 108–112, 116
Smyth, Gerry 18
Spenser, Edmund 157, 168, 158–159
Spoo, Robert 129
Spurr, David 160
suburb 21, 46, 48, 50, 57
– garden suburb 46, 48, 50, 53, 55, 57
Swenarton, Mark 50, 53–54, 66
Symons, Arthur 134
Synge, J. M. 38, 64

T

Tolkowsky, S. 74–75, 78, 96, 79
translation 13, 24, 141, 143–145
trees 7, 20, 23, 48, 74, 78, 101–102, 105–106, 121, 130, 140, 131–132, 134, 136, 142, 147, 157–158, 171–172

– in Brehon Law 131
– wedding in "Cyclops" 134
Tuan, Yi-Fu 59
Turner, Michael 63, 35

U

Unwin, Raymond 49–51, 53, 67, 55, 58
urban environment 19, 58

V

Vico, Giambattista 13, 23, 122–131, 133–135
– Joyce's interest in 137, 110

W

Wall, Eamonn 19, 56
waste 21–22, 25, 69–73, 78–82, 86–94, 108, 110, 172
– reclamation of waste in Ireland 73
– reclamation of waste in Palestine 82
Weizmann, Dr. Chaim 77
Wells, H. G. 22, 87–88, 98
Wenzell, Tim 18
Whiteley, Gillian 71, 95, 72
Wicke, Jennifer 34, 157, 159–160
Williams, Raymond 15, 20, 58, 142
woodkerne 102, 121, 128, 131, 135
Wyndham Land Act 105
Wyse Power, Jennie 119, 121
Wyse Power, John 120, 139, 121

Y

Yeats, W. B. 15, 26, 38, 41, 47–48, 147

Z

Zionism 23, 70, 74, 96, 76, 75, 77–79
Žižek, Slavoj 65, 45

STUDIEN ZU LITERATUR, KULTUR UND UMWELT
STUDIES IN LITERATURE, CULTURE, AND THE ENVIRONMENT

Herausgegeben von / Edited by Hannes Bergthaller, Gabriele Dürbeck, Robert Emmett, Serenella Iovino, Ulrike Plath

Bd. / Vol. 1   Daniel A. Finch-Race/Stephanie Posthumus (eds): French Ecocriticism. From the Early Modern Period to the Twenty-First Century. 2017.

Bd. / Vol. 2   Alessandro Macilenti: Characterising the Anthropocene. Ecological Degradation in Italian Twenty-First Century Literary Writing. 2017.

Bd. / Vol. 3   Gabriele Dürbeck/Christine Kanz/Ralf Zschachlitz (Hrsg.): Ökologischer Wandel in der deutschsprachigen Literatur des 20. und 21. Jahrhunderts – neue Ansätze und Perspektiven. 2018.

Bd. / Vol. 4   Yi-Peng Lai: Eco*Ulysses*. Nature, Nation, Consumption. 2018.

www.peterlang.com